T0238368

SpringerBriefs in Statistics

For further volumes:
http://www.springer.com/series/8921

Mohamed Abdel-Hameed

Lévy Processes and Their Applications in Reliability and Storage

Springer

Mohamed Abdel-Hameed
Department of Statistics
College of Business and Economics
United Arab Emirates University
Al Ain
UAE

ISSN 2191-544X ISSN 2191-5458 (electronic)
ISBN 978-3-642-40074-2 ISBN 978-3-642-40075-9 (eBook)
DOI 10.1007/978-3-642-40075-9
Springer Heidelberg New York Dordrecht London

Library of Congress Control Number: 2013954772

Mathematics Subject Classification (2010): 60K10, 60K20

Printed on acid-free paper

Springer is part of Springer Science+Business Media (www.springer.com)

To

The memory of my parents

Preface

Over the last few decades, Lévy processes have been used extensively in reliability, hydrology, and water resource engineering.

In reliability engineering, they are used to model degradation of devices over time. Certain types of Lévy processes have been found to provide a good model for creep of concrete, fatigue crack growth, corroded steel gates, and chloride ingress into concrete. At the beginning of the work done in reliability, engineers described the uncertainties about the failure times using the survival function; knowing the shape of such a function they can determine and study the properties of the failure rate and, based on that, they can determine the best possible maintenance policies. To estimate the survival function accurately (from a statistical point of view), one has to observe the failure times of many items and these failure random variables are assumed to be independent. In practice, it is not always possible to observe many failures, and even if such failure times are possible to obtain, they are not independent as they all might be affected by an environment. The other approach is to assess the failure of a device based on the characteristics of the process that caused its failure, normally a degradation process. Such an approach is common in assessing the amount of crack, the amount of erosion and creep, and the amount of contamination.

In hydrology and water resource engineering, they are used (among other things) to model the input of water in a reservoir over time. Brownian motion, compound Poisson processes, inverse Gaussian processes, and spectrally positive Lévy processes have been used to describe such input. Knowing the input process and its characteristics enables one to determine and properly improve the cost of running the dam over time.

This monograph consists of three chapters, notations and terminology, and an appendix. In the appendix, we give some basic definitions and results. In Chap. 1 we discuss *Lévy Processes and Their Characteristics*. In Chap. 2, we discuss the applications of Lévy processes in describing *Degradation Processes*. In Chap. 3 we deal with the usage of Lévy processes to describe the input processes and controlling the cost of running reservoirs.

Readers are advised to begin with at least a quick look at the appendix, and the notations and terminology. They serve to review the prerequisite results and definitions. At the end of each chapter as well as in the appendix, relevant references are given. I did not attempt to compile comprehensive bibliographies, but rather give a list of those references that I used to write this book.

Acknowledgments

I want to thank my wife for her patience and understanding. I also thank the external reviewers for their helpful comments on an earlier version of this book. Many thanks to my editor for her insights and many suggestions.

Contents

Notation and Terminology

We let $R = (-\infty, \infty)$, $R_+ = (0, \infty)$, $N = \{1, 2, \ldots\}$, $N_+ = \{0, 1, \ldots\}, \overline{N}_+ = N_+ \cup \{\infty\}$, $\overline{R} = R \cup \{\infty\}$, $\overline{R}_+ = R_+ \cup \{\infty\}$, and $R_0 = R \backslash \{0\}$. We will use the term "increasing" to mean "non-decreasing," and the term "positive" to mean "non-negative." All random variables are defined on a probability space $(\Omega, F, \mathbf{P})$. All processes used have the space of real numbers or subsets of it as their state space. For any process $Y = \{Y_t, t \geq 0\}$, any Borel subset A of the state space and any functional f, $E_y(f)$ denotes the expectation of f conditional on $Y_0 = y$, $P_y(A)$ denotes the corresponding probability measure and $\mathbf{I}_A$ is the indicator function of the set A. In the sequel we will write, indifferently, P_0 or P and E_0 or E. For $x, y \in R$, we define $x \vee y = x \max y$ and $x \wedge y = x \min y$. For every $t \geq 0$, we define $\underline{Y_t} = \inf_{0 \leq s \leq t} (Y_s \wedge 0), \bar{Y}_t = \sup_{0 \leq s \leq t} (Y_s \vee 0)$, For every $t \in R_+$, the process obtained by reflecting Y at its infimum and supremum are defined as follows: $Y_t - \underline{Y_t}$ and $\bar{Y}_t - Y_t$, respectively. For any space G, $\sigma(G)$ denotes the smallest sigma algebra of subsets of G.

The following is a list of symbols that are used in this book.

υ	Lévy measure
ϕ	Laplace Exponent of a spectrally positive Lévy process
$p(t, x, y)$	Probability transition function
$W^{(\alpha)}$	α-scale function
$Z^{(\alpha)}$	Adjoint α-scale function
U^{α}	α-potential measure of a non-subordinator
u^{α}	α-potential density non-subordinator
$\bar{F}$	Survival probability
IFR	Increasing failure rate
DFR	Decreasing failure rate
IFRA	Increasing failure rate average
DFRA	Decreasing failure rate average
TP_2	Totally positive of order 2
Af	Infinitesimal generator
$Q(x, A, t)$	Semi-Markov renewal kernel
$R(x, A, t)$	Renewal function of a Markov renewal process

■	End of proof
$\stackrel{sgn}{=}$	Equal in sign
$\stackrel{def}{=}$	Equal by definition

Chapter 1
Lévy Processes and Their Characteristics

Abstract We give an introductory review of Lévy processes and their properties with emphasis on subordinators and spectrally positive Lévy processes. The α-potentials of these processes are given. Results on the times of first exit of such processes are discussed. Several examples of such processes are given.

Keywords Lévy processes · The Lévy-Itô decomposition formula · Scale functions · Subordinators · Spectrally positive Lévy processes · Killed processes · First exit times · Brownian motion · Gamma processes · Stable processes

1.1 Lévy Processes

Definition 1.1 A stochastic process $X = \{X_t, t \geq 0\}$ is said to be a Lévy process if the following hold:

(i) It has right continuous sample paths with left limits.
(ii) X has *stationary increments*, i.e., for every $s, t \geq 0$, the distribution of $X_{t+s} - X_t$, is independent of t.
(iii) X has independent increments, i.e., for every $t, s \geq 0$, $X_{t+s} - X_t$ is independent of $(X_u, u \leq t)$.

That is to say, a Lévy process is a process with stationary and independent increments whose sample paths are right continuous with left-hand limits.

Any Lévy process X enjoys the following property: For all $t \geq 0$

$$E[e^{i\theta X_t}] = e^{t\Phi(\Theta)}.$$

The function Φ is known as the characteristic function of the process X, it has the form

$$i\theta d - \frac{\theta^2 b}{2} + \int_R [\exp(i\theta x) - 1 - i\theta x I_{\{|x|<1\}}]\nu(dx), \tag{1.1}$$

M. Abdel-Hameed, *Lévy Processes and Their Applications in Reliability and Storage*, SpringerBriefs in Statistics, DOI: 10.1007/978-3-642-40075-9_1, © The Author(s) 2014

where $d \in R$, $b \in R_+$ and ν is a measure on R satisfying $\nu(\{0\}) = 0$, $\int_R (x^2 \wedge 1)\nu(dx) < \infty$.

The measure ν is called the Lévy measure, it characterizes the size and frequency of the jumps. If this measure is infinite, then the process has an infinite number of jumps of every small sizes in any small interval.

Definition 1.2 A Lévy process is said to be of bounded variation if $b = 0$ and $\int_R (| x | \wedge 1)\nu(dx) < \infty$.

For such processes, the characteristic function is of the form

$$\Phi(\Theta) = ia\theta + \int_R [\exp(i\theta x) - 1]\nu(dx), \tag{1.2}$$

where $a = d - \int_{\{|x|<1\}} x\nu(dx)$.

1.2 The Lévy-Itô Decomposition

The Lévy-Itô decomposition identify any Lévy process as the sum of four independent processes, it is stated as follows:

Theorem 1.3 Given any $d \in R, b \in R_+$ and measure ν on R satisfying $\nu(\{0\}) = 0$, $\int_R (1 \wedge x^2)\nu(dx) < \infty$, there exists a probability space $(\Omega, \mathcal{F}, P)$ on which a *Lévy* process X is defined. Furthermore, for each $t \in R_+$

$$X_t = dt + B_t + \int_{[0,t]\times\{|x|>1\}} xM(ds, dx) + \int_{[0,t]\times\{|x|\le 1\}} x(M - m)(ds, dx),$$

where M is a *Poisson random measure* on $R_+ \times R_0$ with mean measure $m(ds, dx) = ds\nu(dx)$ and $\int (x^2 \wedge 1)\nu(dx) < \infty$. Furthermore, $B = \{B_t : t \ge 0\}$ is a Brownian motion with zero mean and variance coefficient b, and d is called the drift term.

The above theorem implies that the jump random measure defined, for any $A \in \sigma(R_+ \times R_0)$, by

$$M(A) = \sum_{s\ge 0} \mathbf{I}_A(s, X_s - X_{s^-}),$$

is a Poisson random measure on $R_+ \times R_0$ with mean measure $m(ds, dx) = ds\nu(dx)$, and $\int (x^2 \wedge 1)\nu(dx) < \infty$. Furthermore, the process X is the sum of four independent processes $\overset{(1)}{X}$, $\overset{(2)}{X}$, $\overset{(3)}{X}$, and $\overset{(4)}{X}$, where $\overset{(1)}{X}$ is a constant drift, $\overset{(2)}{X}$ is a Brownian motion with zero mean and variance coefficient b, $\overset{(3)}{X}$ is a compound Poisson process with arrival rate equal to $\nu(| x |> 1)$, jump magnitude distribution function $F(dx) = \frac{\mathbf{I}_{\{|x|\ge 1\}}\nu(dx)}{\nu(|x|>1)}$, and $\overset{(4)}{X}$ is a pure jump martingale that has countably many jumps over every finite interval; these jumps are of magnitude less than one almost surely. The characteristic exponents of $\overset{(1)}{X}$, $\overset{(2)}{X}$, $\overset{(3)}{X}$, and $\overset{(4)}{X}$ (denoted by $\overset{(1)}{\varphi}, \overset{(2)}{\varphi}, \overset{(3)}{\varphi}$ and $\overset{(4)}{\varphi}$, respectively) are as follows:

$$\overset{(1)}{\varphi}(\theta) = i\theta d$$

$$\overset{(2)}{\varphi}(\theta) = -\frac{\theta^2 b}{2},$$

$$\overset{(3)}{\varphi}(\theta) = \int_{\{|x|\geq 1\}} (\exp(i\theta x) - 1)\nu(dx),$$

$$\overset{(4)}{\varphi}(z) = \int_{\{|x|<1\}} (\exp(i\theta x) - 1 - i\theta x)\nu(dx).$$

1.3 The Strong Markov Property for Lévy Processes

Definition 1.4 For any stopping time T with respect to $\mathcal{F}_\infty$, the sigma algebra generated by T is defined as follows:

$$\mathcal{F}_T = \{A \subset \Omega : A \cap \{T \leq t\} \in \mathcal{F}_t, t \in R_+\}.$$

The next theorem illustrates that the stationarity and the independence of the increments of Lévy processes hold even if the starting time in the increment is a stopping time, instead of being fixed.

Theorem 1.5 Let $L = \{L_t, t \in R_+\}$ be a Lévy processes. For any stopping time T with respect to $\mathcal{F}_\infty$ and for any $t \in R_+$, we define

$$Y_t = L_{T+t} - L_T .$$

Then, on the event $\{T < \infty\}$, the process Y has the same distribution as the process L and is independent of $\mathcal{F}_T$.

Proof Since the indicator function of any event can be approximated by a sequence of bounded continuous function, it suffices to show that, for $m = 1, 2, ..., t_1, ..., t_m \in R_+$, every bounded continuous function $f : R^m :\rightarrow R$, and every $A \in \mathcal{F}_T$

$$E[\mathbf{I}_{A\cap\{T<\infty\}} f(Y_{t_1}, .., Y_{t_m})] = P(A \cap \{T < \infty\})E[f(L_{t_1}, .., L_{t_m})].$$

From Theorem 10 in the Appendix, it suffices to show that the above identity holds for $m = 2$, and $f = f_1 f_2$ where f_1 , $f_2 : R :\rightarrow R$ are bounded continuous functions

For every $n = 1, 2, \ldots$, we define

$$T^n = \sum_{k\geq 1} \frac{k}{2^n} \mathbf{I}_{\{\frac{k-1}{2^n} < T \leq \frac{k}{2^n}\}}.$$

For every $n = 1, 2, \ldots$, $t \in R_+$, we define $Y_t^n = L_{T^n+t} - L_{T^n}$. It is known that $T^n \downarrow T$, as $n \uparrow \infty$ almost surely. Since the process L is right continuous, f_1 and f_2 are continuous functions, it follows that $f_i(Y_{t_i}^n) \to f_i(Y_{t_i})$ almost surely, as $n \to \infty$, $i = 1, 2$. Using the *bounded convergence theorem* we have

$$\lim_{n\to\infty} E[\mathbf{I}_{A\cap\{T^n<\infty\}} f_1(Y_{t_1}^n) f_2(Y_{t_2}^n)] = E[\mathbf{I}_{A\cap\{T<\infty\}} f_1(Y_{t_1}) f_2(Y_{t_2})]$$

For simplicity, for k = 1,2,..., we will denote the set $\{\frac{k-1}{2^n} < T \leq \frac{k}{2^n}\}$ by A_k. For $A \in \mathcal{F}_T$, we write

$$\begin{aligned}
E[\mathbf{I}_{A\cap\{T^n<\infty\}} & f_1(Y_{t_1}^n) f_2(Y_{t_2}^n)] \\
&= E[\sum_{k\geq 1} \mathbf{I}_{A\cap A_k} f_1(L_{\frac{k}{2^n}+t_1} - L_{\frac{k}{2^n}}) f_2(L_{\frac{k}{2^n}+t_2} - L_{\frac{k}{2^n}})] \\
&= E[E[\sum_{k\geq 1} \mathbf{I}_{A\cap A_k} f_1(L_{\frac{k}{2^n}+t_1} - L_{\frac{k}{2^n}}) f_2(L_{\frac{k}{2^n}+t_2} - L_{\frac{k}{2^n}}) \mid \mathcal{F}_{\frac{k}{2^n}}]] \\
&= E[\sum_{k\geq 1} \mathbf{I}_{A\cap A_k} E[f_1(L_{t_1}) f_2(L_{t_2})]] \\
&= P(A \cap \{T^n < \infty\}) E[f_1(L_{t_1}) f_2(L_{t_2})],
\end{aligned}$$

where the third equation follows since the process L has stationary independent increments and since, for $k = 1, 2, \ldots$, and $A \in \mathcal{F}_T$ the event $A \cap A_k \in \mathcal{F}_{\frac{k}{2^n}}$. Our assertion is proved by letting $n \to \infty$, in both sides of the last equation above. ■

The following shows that every Lévy process is a strong Markov process in the sense described in the Appendix.

Corollary 1.6 Let L be a Lévy process, then L is a strong Markov process.

Proof Let T be an arbitrary stopping time, we need to show that, for each $t \in R_+$ and every bounded function f,

$$E[f(L_{t+T}) \mid \mathcal{F}_T] = E[f(L_{t+T}) \mid L_T].$$

Let the process $Y = \{Y_t, t \in R_+\}$ be as defined in the previous theorem. Observe that $L_{t+T} = Y_t + L_T$, hence $E[f(L_{t+T}) \mid \mathcal{F}_T] = E[f(Y_t + L_T)) \mid \mathcal{F}_T]$. From the above theorem Y_t has the same distribution as the L_t and is independent of $\mathcal{F}_T$. Thus, $E[f(L_{t+T}) \mid \mathcal{F}_T] = E[f(Y_t + L_T)) \mid L_T] = E[f(L_t + L_T)) \mid L_T]$, L_t is independent fro L_T. But, given L_T, the random variable $L_t + L_T$ has the same distribution as L_{t+T} and our assertion is proven. ■

1.4 Subordinators

Definition 1.7 A Lévy process is called a *subordinator* if its sample paths are increasing.

From Theorem 1.3 it follows that, for each $t \in R_+$

$$X_t = \zeta t + \int_{[0,t)\times R_+} xM(ds, dx),$$

where $\zeta \geq 0$, and M is a Poisson random measure on $R_+ \times R_+$ with mean measure $m(ds, dx) = ds\upsilon(dx)$, and $\int_0^\infty (x \wedge 1)\upsilon(dx) < \infty$.

For such processes we have, for all $\theta \geq 0$,

$$E[e^{-\theta X_t}] = e^{-t\psi(\theta)}, \tag{1.3}$$

where

$$\psi(\theta) = \zeta\theta + \int_0^\infty (1 - e^{-\theta x})\upsilon(dx),$$

and $\zeta \geq 0$ is the drift term.

The function ψ is called the *Laplace exponent* of the subordinator. It follows that every subordinator is of bounded variation.

We now mention some examples of subordinators.

Example 1 *Compound Poisson processes.* A subordinator X with finite Lévy measure is called a *compound Poisson process with a positive drift.* In this case, $\upsilon(dx) = \lambda F(dx)$, where $\lambda > 0$, F is the distribution function with support R_+, and the corresponding Poisson random measure M is finite. Let $T = (T_n, n = 1, 2, ...)$ be a sequence of independent identically distributed exponential random variables, with mean $\frac{1}{\lambda}$. For $n \in N_+$, let $S_n = T_1 + \cdots + T_n$, $S_0 \equiv 0$, then, for every $t \in R_+$

$$X_t = \zeta t + \sum_n X_n \mathbf{I}_{(0,t]}(S_n)$$

where $\{X_n, n = 1, 2, ...\}$ is a sequence of independent positive identically distributed random variables with distribution function F, and independent of T.

In this case,

$$\psi(\theta) = \theta\zeta + \lambda \int_0^\infty (1 - e^{-\theta x})F(dx). \tag{1.4}$$

Assuming $\zeta = 0$, and that F has a density f, then the probability transition function of this process is as follows:

$$p(t,x,y) = \begin{cases} \sum_{n=0}^{\infty} e^{-\lambda t}\frac{(\lambda t)^n}{n!} f^{(n)}(y-x), & y > x \\ 0, & y \le x, \end{cases}$$

where $f^{(n)}$ is the nth convolution of f with itself.

Example 2 *Inverse Brownian motion.* A subordinator X with Lévy measure of the form

$$\upsilon(dx) = \frac{1}{\sqrt{2\pi\sigma^2 x^3}} \exp\left(\frac{-x\mu^2}{2\sigma^2}\right).$$

is called *inverse Brownian process.*

In this case,

$$\psi(\theta) = \theta\zeta + \frac{1}{\sigma^2}(\sqrt{2\theta\sigma^2 + \mu^2} - \mu). \tag{1.5}$$

If $\zeta = 0$, then, *for* $t,x,y \in R_+$, the probability transition function of the process X is as follows:

$$p(t,x,y) = \begin{cases} \frac{t}{\sigma\sqrt{2\pi(y-x)^3}} \exp\{-\frac{[\mu(y-x)-t]^2}{2(y-x)\sigma^2}\}, & y > x \\ 0, & y \le x. \end{cases}$$

In this case, $EX_1 = \frac{1}{\mu}$, and $Var(X_1) = \frac{\sigma^2}{\mu^3}$.

Example 3 *Gamma processes.* A subordinator X with Lévy measure

$$\nu(dx) = \frac{\alpha}{x}\exp(-x\beta)dx, \qquad x > 0$$

where $\alpha, \beta > 0$, is called a *gamma process.*

It follows that

$$\psi(\theta) = \theta\zeta + \alpha \ln(1 + \theta/\beta). \tag{1.6}$$

Furthermore, if $\zeta = 0$, its probability transition function is of the form

$$p(t,x,y) = \begin{cases} \frac{\beta^{at}}{\Gamma(at)} e^{-(y-x)\beta} (y-x)^{at-1}, & y > x \\ 0, & y \le x. \end{cases}$$

In this case, the mean term $(E(X_1))$ and the variance term $(V(X_1))$ are equal to α/β and α/β^2, respectively.

Example 4 *Stable processes with stability parameter* β, $\beta \in (0, 1)$. A subordinator X with Lévy measure

$$\nu(dx) = \frac{\beta}{\Gamma(1-\beta)x^{\beta+1}} \qquad x > 0$$

is called a stable process with stability parameter β.

In this case

$$\psi(\theta) = \theta\zeta + \theta^{\beta}, \tag{1.7}$$

and $E[X_t] = \infty$, for all $t \geq 0$.

We will discuss stable processes in general and the case when the index $\beta \in (1, 2)$ in Sect. 1.6 of this chapter.

Example 5 *Generalized stable processes with stability parameter* β, $\beta \in (0, 1)$. A subordinator X with Lévy measure

$$\nu(dx) = \frac{\beta e^{-\lambda x}}{\Gamma(1-\beta)x^{\beta+1}}dx \qquad x, \lambda > 0$$

is called a generalized stable subordinator.

It is easily seen that

$$\psi(\theta) = \theta\zeta + (\theta+\lambda)^{\beta} - \lambda^{\beta}, \tag{1.8}$$

and $E[X_1] = \zeta + \beta\lambda^{\beta-1}$.

1.5 Spectrally Positive Processes

Definition 1.8 A non-subordinator is said to be spectrally positive (negative) if it has no negative (positive) jumps.

For any spectrally positive process L, we let $\hat{L} = -L$, throughout. It is clear that L is spectrally positive if and only if the process $\hat{L}$ is spectrally negative.

From (1.1), it follows that, for each $\theta \in R_+$, the Laplace transform $E[e^{-\theta L_t})]$ exists, furthermore,

$$E[e^{-\theta L_t})] = e^{t\phi(\theta)},$$

where

$$\phi(\theta) = -d\theta + \frac{\theta^2\sigma^2}{2} - \int_0^{\infty}(1 - e^{-\theta x} - \theta x 1_{\{x<1\}})\upsilon(dx). \tag{1.9}$$

The term $d \in R$ is the drift term, $\sigma^2 \in R_+$ is the variance of the Brownian motion and υ is a positive measure on $[0, \infty)$, $\upsilon(\{0\}) = 0$, and $\int_0^{\infty}(x^2 \wedge 1)\nu(dx) < \infty$.

The function ϕ is known as the *Laplace exponent* of the spectrally positive process. The following gives some properties of the Laplace component above:

Lemma 1.9 Let ϕ be as defined in (1.9). Then

(i) $\phi(0) = 0$.
(ii) ϕ is a convex function in its argument.
(iii) If $\phi^{'}(0+) > 0$, then ϕ is strictly increasing on R_+.
(iv) If $\phi^{'}(0+) \leq 0$, then there exists $\theta^* > 0$ such that $\phi(\theta) < 0$ if $\theta < \theta^*$, and $\phi(\theta) \geq 0$ and increasing if $\theta \geq \theta^*$.
(vi) $\lim_{\theta \to \infty} \phi(\theta) = \infty$.

Proof (i) This follows immediately from the definition of ϕ.
(ii) From the definition of ϕ, it follows that $\phi^{''}(\theta)$ has the same sign as $E[e^{-\theta L_1}]$ $E[L_1^2 e^{-\theta L_1}] - (E[L_1 e^{-\theta L_1}])^2$. The fact that this term is positive, is easily seen from Hölder inequality. This establishes the assertion.
(iii) This assertion also follows from (ii) above.
(iv) This assertion also follows from (i) and (ii) above.
(vi) Since the process L is spectrally positive, then there exists a $t \in (0, \infty)$ such that $P\{L_t < 0\} > 0$. For such a t, $e^{t\phi(\theta)} = E[e^{-\theta L_t}] \geq E[e^{-\theta L_t}, L_t < 0]$. The assertion follows by letting $\theta \to \infty$, in the last inequality. ■

It is clear that, $\phi^{'}(0+) = -E[L_1]$. For $\alpha \in R_+$, we define $\eta(\alpha) = \phi^{-1}(\alpha)$, i.e,

$$\eta(\alpha) = \sup\{\theta : \phi(\theta) = \alpha\}, \tag{1.10}$$

It is seen that $\eta(0) = 0$ if and only $E[L_1] \leq 0$. Note that, $E(L_1) = \int_1^\infty x\upsilon(dx) + d$. Furthermore, $\lim_{t\to\infty} L_t = \infty$ if and only if $E(L_1) > 0$, and $\lim_{t\to\infty} L_t = -\infty$ if and only if $E[L_1] < 0$. Also, if $E(L_1) = 0$, then L oscillates from $-\infty$ to ∞.

A version of the following theorem is given in Theorem 1 of [1], it is also included in Theorem 8 p. 194 of [2].

Theorem 1.10 Let X be a spectrally positive process, with Laplace exponent ϕ, and η is as defined in (1.10). Then, there exists an absolutely continuous increasing function W such that,

$$\int_0^\infty e^{-\theta x} W(x)dx = \frac{1}{\phi(\theta)}, \theta > \eta(0). \tag{1.11}$$

Definition 1.11 For any spectrally positive process with Laplace component ϕ and for $\alpha \geq 0$, the α-scale function W^α: $R \twoheadrightarrow R_+$, $W^\alpha(x) = 0$ for every $x < 0$, and on $[0, \infty)$ it is defined as the unique continuous increasing function such that

$$\int_0^\infty e^{-\theta x} W^{(\alpha)}(x)dx = \frac{1}{\phi(\theta) - \alpha}, \theta > \eta(\alpha). \tag{1.12}$$

The existence of $W^{(\alpha)}$ and its relation to W above is established as follows. Since $\theta > \eta(\alpha)$ if and only if $\phi(\theta) > \alpha$, then we have

$$\begin{aligned}
\frac{1}{\phi(\theta)-\alpha} &= \frac{1}{\phi(\theta)}[\frac{1}{1-\alpha/\phi(\theta)}] \\
&= \sum_{k\geq 0} \alpha^k [\frac{1}{\phi(\theta)}]^{k+1} \\
&= \sum_{k\geq 0} \alpha^k [\int_0^\infty e^{-\theta x} W(x)]^{k+1} dx \\
&= \sum_{k\geq 0} \alpha^k \int_0^\infty e^{-\theta x} W^{*(k+1)}(x) dx,
\end{aligned}$$

where for k = 1, 2, ..., $W^{*(k)}$ is the kth convolution of W with itself. Note that, since W is increasing

$$\begin{aligned}
W^{*(2)}(x) &= \int_0^x W(x-y)W(y)dy \\
&\leq \frac{x}{1!} W(x)^2.
\end{aligned}$$

By induction on k, it follows that for $k \geq 1$,

$$W^{*(k+1)}(x) \leq \frac{x^k}{k!}(W(x))^{k+1}.$$

Hence, for each $x \in R_+$, the series $\sum_{k\geq 0} \alpha^k W^{*(k+1)}(x)$ converges. Using Fubini's Theorem we have

$$\sum_{k\geq 0} \alpha^k \int_0^\infty e^{-\theta x} W^{*(k+1)}(x) dx = \int_0^\infty e^{-\theta x} \sum_{k\geq 0} \alpha^k W^{*(k+1)}(x) dx.$$

From the uniqueness of the Laplace transform, we have, for $\alpha > 0$

$$W^{(\alpha)}(x) = \sum_{k=0}^\infty \alpha^k W^{*(k+1)}(x). \tag{1.13}$$

If a spectrally positive Lévy process has bounded variation, then using (1.2) it follows that

$$\phi(\theta) = \zeta\theta - \int_0^\infty (1-e^{-\theta x})\upsilon(dx). \tag{1.14}$$

where

$$\zeta = \int_{\{|x|<1\}} x\nu(dx) - d > 0. \tag{1.15}$$

In this case, we can write, for each $t \geq 0$

$$X_t = Y_t - \zeta t$$

where the process Y is a subordinator with drift term equal to zero,

Lemma 1.12 Let X be a spectrally positive process. Then, for each $\alpha > 0$

(a) $W^{(\alpha)}(0) = \frac{1}{\zeta}$ if and only if X is of bounded variation, where ζ is given in (1.15).
(b) $W^{(\alpha)}(0) = 0$ if and only if X is of unbounded variation.

Proof (a) From the *initial value theorem* for the Laplace transform, and (1.12) we have

$$\begin{aligned} W^{(\alpha)}(0) &= \lim_{\theta\to\infty} \int_0^\infty \theta e^{-\theta x} W^{(\alpha)}(x)dx \\ &= \lim_{\theta\to\infty} \frac{\theta}{\phi(\theta) - \alpha} \\ &= (\lim_{\theta\to\infty} \frac{\phi(\theta)}{\theta})^{-1}. \end{aligned}$$

Since, for $x, \theta \in R_+$ and θ large enough, $(1 - e^{-\theta x}) \leq (\theta x \wedge 1) < \theta(x \wedge 1)$, using the fact that $\int_0^\infty (x \wedge 1)\upsilon(dx) < \infty$, (1.14) and the *Lebesgue dominated convergence theorem*, we have

$$W^{(\alpha)}(0) = \frac{1}{\zeta}$$

if and only if X is of bounded variation.

(b) The assertion that $W^{(\alpha)}(0) = 0$ if the process L is of unbounded variation follows, since in this case and from the definition of ϕ, $\lim_{\theta\to\infty} \frac{\theta}{\phi(\theta)-\alpha} = 0$. ■

Furthermore, (see Lemma 8.2 of [3]), $W^{(\alpha)}$ is right and left differentiable on $(0, \infty)$. By $W_+^{(\alpha)'}(x)$, we will denote the right derivative of $W^{(\alpha)}$ in x.

The adjoint α-scale function associated with $W^{(\alpha)}$ (denoted by $Z^{(\alpha)}$) is defined as follows:

Definition 1.13 For $\alpha \geq 0$, the *adjoint* α-scale function $Z^{(\alpha)} : R_+ \twoheadrightarrow [1, \infty)$ is defined as

$$Z^{(\alpha)}(x) = 1 + \alpha \int_0^x W^{(\alpha)}(y)dy. \tag{1.16}$$

Lemma 1.14 For $\alpha > 0$

(a)

$$W^{(\alpha)}(x) \sim \frac{e^{\eta(\alpha)\,x}}{\phi'(\eta(\alpha)\,)}\ , \text{ as } x \to \infty. \tag{1.17}$$

(b)

$$Z^{(\alpha)}(x) \sim \frac{\alpha e^{\eta(\alpha)\,x}}{\eta(\alpha)\phi'(\eta(\alpha)\,)}\ , \text{ as } x \to \infty. \tag{1.18}$$

Proof (a) Let $\overset{*}{W}{}^{(\alpha)}(x) = e^{-\eta(\alpha)x} W^{(\alpha)}(x)$, then from (1.12) we have, for $\theta \in R_+$

$$\int_0^\infty e^{-\theta x} \overset{*}{W}{}^{(\alpha)}(x)dx = \frac{1}{\phi(\theta + \eta(\alpha)) - \alpha}.$$

From the *final-value theorem* of the Laplace transform we have

$$\begin{aligned}
\lim_{x\to\infty} \overset{*}{W}{}^{(\alpha)}(x) &= \lim_{\theta\to 0} \int_0^\infty \theta e^{-\theta x} \overset{*}{W}{}^{(\alpha)}(x)dx \\
&= \lim_{\theta\to 0} \frac{\theta}{\phi(\theta + \eta(\alpha)) - \alpha} \\
&= \lim_{\theta\to 0} \frac{\theta}{\phi(\theta + \eta(\alpha)) - \phi(\eta(\alpha))} \\
&= \frac{1}{\phi'(\eta(\alpha))}.
\end{aligned}$$

Hence,

$$W^{(\alpha)}(x) \sim \frac{e^{\eta(\alpha)\,x}}{\phi'(\eta(\alpha)\,)}, \text{ as } x \to \infty.$$

(b) From (1.16) and (1.17), it follows that as $x \to \infty$, for $\alpha > 0$, $\frac{Z^{(\alpha)}(x)}{W^{(\alpha)}(x)} \sim \frac{\alpha W^{(\alpha)}(x)}{W'^{(\alpha)}(x)} = \frac{\alpha}{\eta(\alpha)}$, hence

$$Z^{(\alpha)}(x) \sim \frac{\alpha e^{\eta(\alpha)\,x}}{\eta(\alpha)\phi'(\eta(\alpha)\,)}, \text{ as } x \to \infty.$$

1.6 Examples of Spectrally Positive Processes

Example 1 *Brownian Motion*. The Brownian motion with mean $\mu \in R$, variance term σ^2, is an example of spectrally positive Lévy processes, where $\nu(R_+) = 0$. From (1.1) we have, that for $\theta \geq 0$, $\phi(\theta) = -\mu\theta + \frac{\theta^2\sigma^2}{2}$. It follows that, for $\alpha \geq 0$, $\eta(\alpha) = \frac{\sqrt{2\alpha\sigma^2+\mu^2}+\mu}{\sigma^2}$. For each $t \in R_+, x, y \in R$, the transition probability function of this process is given as follows:

$$p(t, x, y) = \frac{1}{\sqrt{2\pi b^2 t}} \exp\{\frac{(y - x - \mu t)^2}{2\sigma^2}\}.$$

Let $\delta = \sqrt{2\alpha\sigma^2 + \mu^2}$, then

$$W^{(\alpha)}(x) = \frac{2}{\delta} e^{\mu x/\sigma^2} \sinh(x\delta/\sigma^2),$$
$$Z^{(\alpha)}(x) = e^{\mu x/\sigma^2}(\cosh(x\delta/\sigma^2) - \frac{\mu}{\delta}\sinh(x\delta/\sigma^2)) \tag{1.19}$$

Example 2 *Stable processes with stability parameter* $\beta \in (1, 2)$. A Lévy process X is called stable process with stability parameter $\beta > 0$, if its Lévy measure has support $[0, \infty)$ and for each $t \geq 0$, X_t has the same distribution as $t^{(1/\beta)} X_1$. When $\beta \in (0, 1)$, the process X is a subordinator with no drift, as discussed in Example 4 of Sect. 1.4. Here we will deal with the case where $\beta \in (1, 2)$, in this case the process is spectrally positive. Let X be such a process, it follows that for $t, \theta \geq 0$,

$$E[e^{-\theta X_t}] = E[e^{-\theta t^{(1/\beta)} X_1}].$$

Since the left-hand side of the above equation is equal to $e^{t\phi(\theta)}$, then we must have

$$E[e^{-\theta t^{(1/\beta)} X_1}] = e^{t\phi(\theta)}.$$

Clearly $\phi(\theta) = C\theta^\beta$, is the solution of the last equation. Since $\lim_{\infty\to\infty} \phi(\theta) = \infty$, (Lemma 1.9 (vi)), the constant C must be greater than zero. In summary

$$\phi(\theta) = C\theta^\beta, \tag{1.20}$$

$C > 0$. In this case, the Lévy measure is of the form

$$\nu(dx) = \frac{a}{x^{\beta+1}}, \tag{1.21}$$

where a is a positive real number.

It follows that, for all $t \geq 0$, $E[X_t] = 0$ and the value of the term d in (1.9) is equal $-\int_1^{\infty} x\upsilon(dx)$. Furthermore,

$$\phi(\theta) = \int_0^{\infty} (e^{-\theta x} - 1 + \theta x)\upsilon(dx). \tag{1.22}$$

If C in (1.20) is taken to be 1, then the constant a in (1.21) is found to be $\frac{1}{\Gamma(-\beta)}$. In this case,

$$Z^{(\alpha)}(x) = E_{\beta}(\alpha x^{\beta}) \tag{1.23}$$

$$W^{(\alpha)}(x) = \beta x^{\beta-1} E'_{\beta}(\alpha x^{\beta}), \tag{1.24}$$

where, for $\upsilon > 0$, $E_{\upsilon}(x) = \sum_{k\geq 0} x^k / \Gamma(1 + \upsilon k)$ is the Mittag–Leffler function. (see [3], p. 233)

The process X jumps upwards only and creeps downwards (in the sense that, for every negative x, $P\{X_{T_x^-} = x\} = 1$, where T_x^- is the first time the process X hits x from above). Furthermore, $\sigma = 0$, and $\int_0^{\infty} (x \wedge 1)\upsilon(dx) = \infty$, thus X is of unbounded variation.

Example 3 *Spectrally positive processes of bounded variation.* Assume that X is a spectrally positive process of bounded variation, with Laplace exponent given in (1.14). Let $\mu = \int_0^{\infty} x\upsilon(dx)$ and assume that $\mu < \infty$. For every $x \in R_+$, we let $\bar{\upsilon}(x) = \upsilon((x, \infty))$, define the probability density function $f(x) = \frac{\bar{\upsilon}(x))}{\mu}$, $F(x)$ as the distribution function corresponding to f, and $\rho = \frac{\mu}{\varsigma}$ which is assumed to be ρ less than one. From (1.14), we have

$$\begin{aligned} \phi(\theta) &= \zeta\theta - \int_0^{\infty} (1 - e^{-\theta x})\upsilon(dx) \\ &= \zeta\theta - \theta \int_0^{\infty} e^{-\theta x} \bar{\upsilon}(x)dx. \end{aligned}$$

Thus,

$$\begin{aligned} \frac{1}{\phi(\theta)} &= \frac{1}{\varsigma\theta[1 - \rho \int_0^{\infty} e^{-\theta x} f(x)dx]} \\ &= \frac{1}{\varsigma\theta} \int_0^{\infty} e^{-\theta x} \sum_{n=0}^{\infty} \rho^n f^{(n)}(x)dx \\ &= \frac{1}{\varsigma} \int_0^{\infty} e^{-\theta x} \sum_{n=0}^{\infty} \rho^n F^{(n)}(x)dx. \end{aligned}$$

Since, $\rho < 1$, then $\eta(0) = 0$, thus

$$\int_0^{\infty} e^{-\theta x} W(x)dx = \frac{1}{\phi(\theta)}, \theta > 0.$$

Thus, we must have

$$W(x) = \frac{1}{\varsigma} \sum_{n=0}^{\infty} \rho^n F^{(n)}(x). \tag{1.25}$$

The following are three examples of spectrally positive processes of bounded variation.

Example 4 *Spectrally positive processes of bounded variation with a gamma subordinator.* If X is a spectrally positive process with gamma subordinator, then from (1.14), for each $t \geq 0$,

$$X_t = Y_t - \zeta t,$$

$\zeta > 0$, and the process Y is a gamma process with drift term equal to zero, and parameters $\alpha, \beta > 0$, in the sense described in Example 3 of Sect. 1.4.

In this case the Laplace exponent of the process X is given as follows:

$$\phi(\theta) = \zeta\theta - \alpha \ln(1 + \theta\beta). \tag{1.26}$$

Note that, $\mu = E[Y_1] = \alpha\beta < \infty$. Then, assuming that $\alpha\beta < \zeta$, the scale function W is computed using (1.25), where $\rho = \frac{\alpha\beta}{\varsigma}$, and, for $x > 0$, $\bar{\upsilon}(x) = \int\limits_x^{\infty} \frac{\alpha}{y} exp(-y/\beta)dy$, $F(x) = \int_{[0,x)} \bar{\upsilon}(y)dy/\alpha\beta$.

Example 5 *Spectrally positive processes of bounded variation with a stable subordinator.* From Example 4 of Sect. 1.4, for $\beta \in (0, 1)$, it follows that

$$\phi(\theta) = \theta\zeta - \theta^{\beta}, \tag{1.27}$$

$\zeta > 0$.

In this case $\eta(0) = \zeta^{(\frac{1}{\beta-1})}$, $\mu = \infty$. Thus, we cannot apply (1.25) to compute the scale function. However, when $\zeta = 1$, then from (1.11) and (1.27), we have

$$\int_0^{\infty} e^{-\theta x} W(x)dx = \frac{1}{\zeta\theta - \theta^{\beta}}, \quad \theta > \zeta^{(\frac{1}{\beta-1})}.$$

It can be shown that the solution of the last equation above is

$$W(x) = \frac{1}{\zeta} E_{1-\beta}(\frac{x^{1-\beta}}{\zeta}), \tag{1.28}$$

where, for $v > 0$, $E_v(x)$ is the Mittag–Leffler function with parameter v, which is defined in Example 2 of this section.

Example 6 *Spectrally positive processes of bounded variation with a generalized stable subordinator. Let X* be a spectrally positive process, with generalized stable subordinator. From Example 5 of Sect. 1.4 and (1.14), the Laplace exponent of X is of the form

$$\phi(\theta) = \theta\zeta - (\theta + \lambda)^{\beta} + \lambda^{\beta}, \tag{1.29}$$

where $\beta \in (0, 1)$, $\zeta > 0$, and $\lambda > 0$.

In this case, $\mu = \lambda^{\beta-1} < \infty$. Assuming that $\lambda^{\beta-1} < \zeta$, we can use (1.25) to compute the scale function W, with the following ingredients: $\rho = \frac{\lambda^{\beta-1}}{\zeta}$ and $F'(x) = \frac{\lambda^{1-\beta}}{\Gamma(1-\beta)} \int_x^{\infty} \frac{e^{-\lambda y}}{y^{\beta+1}} dy$.

1.7 The Compensation Formula

For a proof of the following theorem the reader should consult [4], also see Chapter II of [5].

Theorem 1.15 Let X be a Lévy process, defined on a probability space (Ω, P). Let M be a random measure on $(R_+ \times R_0)$. Then, M is a Poisson random measure with mean measure $ds\nu(dx)$ if and only if

$$E[\int_{[0,t)\times R_0} G_s f(x) M(ds, dx)] = E[\int_{[0,t)\times R_0} G_s f(x) ds\nu(dx)],$$

for each $t \in R_+$, for every positive measurable function f on R_0, and every $\mathcal{F}_t$ predictable process (G_s).

The following is an extension of the above theorem.

Theorem 1.16 Let $g(t, x, \omega)$ be such that
(i) $x \to g(t, x, \omega)$ is a positive bounded measurable function, and
(ii) $t \to g(t, x, \omega)$ is predictable with respect to $\mathcal{F}_t$.
Then, For each $t \in R_+$, we have

$$E\left(\int_{[0,t)\times R_0} g(s,x)M(ds,dx)\right) = E\left(\int_{[0,t)\times R_0} g(s,x)ds\nu(dx)\right)$$

Proof We use the monotone class theorem. Take $\mathcal{F}$ to be the class of functions for which the above equation holds. Let $Ł = \{g(s,x) : g(s,x) = G_s f,\ \text{where} f : R \to R_+$ is a measurable function, and (G_s) is F_t − predictable$\}$. From Theorem 1.15, we have $\mathcal{F} \supset Ł$. It is clear that $\mathcal{F}$ is a vector space that contains the constant functions and, by the *monotone convergence theorem,* is closed under taking monotone limits of functions. From Theorem 10 of the appendix $\mathcal{F}$ contains every bounded $\sigma(Ł)$ measurable function. But $\sigma(Ł)$ is nothing but the sigma algebra generated by functions satisfying conditions (i) and (ii) of this theorem. Thus the class of all functions g satisfying the assumptions of this theorem are in $\mathcal{F}$, this finishes the proof. ■

1.8 Non-homogeneous Lévy Processes

The classes of Lévy processes dealt with thus far are known as "homogeneous Lévy processes". Nonhomogeneous Lévy processes are encountered in practice. More than one definition of such processes are found in the literature. The following definition of such processes is suitable for our purposes.

A nonhomogeneous subordinator has the same properties as the homogeneous subordinator with the exception that the increments are not stationary. In this case, we have that, for each $t, \theta \geq 0$

$$E[e^{-\theta X_t}] = \exp\left(-\theta\Lambda(t) - \int_{[0,t]\times R_+} (1-e^{-\theta x})n(ds,dx)\right), \tag{1.30}$$

where $n(ds,dx) = \Lambda(ds)\nu(dx)$, $\int_0^\infty (x \wedge 1)\upsilon(dx) < \infty$, Λ is an arbitrary positive measure on R_+ with $0 \leq \Lambda[0,t] < \infty$ for every $t \geq 0$, and $\Lambda(0) = 0$. We assume that the function $t \to \Lambda(t) \equiv \Lambda[0,t]$ is continuous. It follows that a stochastic process X is a nonhomogeneous subordinator, if and only if, for every $t \in R_+$, $X_t = Y_{\Lambda(t)}$, where the process Y is a homogeneous subordinator, and Λ is as defined above.

In the same manner we define a nonhomogeneous Lévy process as a stochastic process L, for every $t \in R_+$, $L_t = Y_{\Lambda(t)}$, where Y is homogeneous Lévy process, and Λ is as defined above. In this case we have

$$E[e^{i\theta L_t}] = \exp\left(\begin{array}{c}\int_{[0,t]\times R}[\exp(i\theta x) - 1 - i\theta x I_{\{|x|<1\}}]\Lambda(ds)\upsilon(dx)\\ +i\theta a\Lambda(t) - \frac{\theta^2 b}{2}\Lambda(t)\end{array}\right), \tag{1.31}$$

where $\int_{R_0}(x^2 \wedge 1)\upsilon(dx) < \infty$.

1.9 Potentials

We begin by defining the α-potential measure $\mathbf{R}^{\alpha}(x, .)$.

Definition 1.17 Let X be a stochastic process with state space S. For $x \in S$, any Borel set $A \subset S$, and $\alpha \geq 0$

$$\mathbf{R}^{\alpha}(x, A) = E_x \int_0^{\infty} e^{-\alpha t} \mathbf{I}_{\{X_t \in A\}} dt = \int_0^{\infty} P_t(x, A) e^{-\alpha t} dt. \tag{1.32}$$

Since every bounded measurable function can be approximated by a sequence of simple functions, from the *bounded convergence theorem,* and (1.32) it follows that for every bounded measurable function f on S

$$\mathbf{R}^{\alpha} f(x) = E_x \int_0^{\infty} e^{-\alpha t} f(X_t) dt = \int_0^{\infty} f(y) \mathbf{R}^{\alpha}(x, dy). \tag{1.33}$$

We note that if X is a Lèvy process, then $\mathbf{R}^{\alpha}(x, dy) = \mathbf{R}^{\alpha}(0, dy - x)$. We will denote $\mathbf{R}^{\alpha}(0, dy)$ by $\mathbf{R}^{\alpha}(dy)$ throughout.

Lemma 1.18 Let X be a subordinator, with Laplace exponent ψ given in (1.3), then $\mathbf{R}^{\alpha}(dy)$ is obtained by inverting the function $\frac{1}{\alpha + \phi(\theta)}$ with respect to θ.

Proof For $\theta > 0$, let $f(x) = e^{-\theta x}$, $x > 0$, in (1.33). Then,

$$\begin{aligned}
\mathbf{R}^{\alpha} f(0) &= E \int_0^{\infty} e^{-\alpha t} e^{-\theta X_t} dt \\
&= \int_0^{\infty} e^{-\alpha t} E[e^{-\theta X_t}] dt \\
&= \int_0^{\infty} e^{-\alpha t} e^{-t\phi(\theta)} dt \\
&= \frac{1}{\alpha + \psi(\theta)},
\end{aligned}$$

where the second equation above follows from *Fubini's theorem*. The assertion follows, since for $f(x) = e^{-\theta x}$, $\mathbf{R}^{\alpha} f(0) = \int_0^{\infty} e^{-\theta y} \mathbf{R}^{\alpha}(dy)$. ■

Corollary 1.19 Let X be a compound Poisson process with no drift, rate λ, and jump distribution function F whose support is R_+. For $\alpha \geq 0$, let $F_{\alpha} = \frac{\lambda}{\lambda + \alpha} F$, for $n =, 1, \ldots,$ $F_{\alpha}^{(n)}$ is the nth convolution of F_{α}, $F^{(0)}$ is the Dirac measure $\delta_0(x)$, and we write $F_{\alpha}^{(n)}(dy)$ instead of $dF_{\alpha}^{(n)}(y)$. Then, for each $y \geq 0$,

$$\mathbf{R}^{\alpha}(dy) = \frac{1}{(\alpha + \lambda)} \sum_{n \geq 0} F_{\alpha}^{(n)}(dy). \tag{1.34}$$

Proof Let the function f be as defined in the proof of Lemma 1.18. Note that $\alpha - \psi(\theta) = \alpha + \lambda \int_0^\infty (1 - e^{-\theta x}) F(dx) = \alpha + \lambda - \lambda \int_0^\infty e^{-\theta x} F(dx) = (\alpha + \lambda)(1 - \int_0^\infty e^{-\theta x} F_\alpha(dx))$. Thus,

$$\mathbf{R}^\alpha f(0) = \frac{1}{\alpha + \psi(\theta)}$$
$$= \frac{1}{(\alpha + \lambda)} \frac{1}{(1 - \int_0^\infty e^{-\theta x} F_\alpha(dx))}$$

The result is immediate from Lemma 1.18 upon inverting the right-hand side of the last equation with respect to θ. ■

Corollary 1.20 Assume that X is an inverse Gaussian process, as defined in Example 2 of Sect. 1.4. Let φ be the density function of the standard normal random variable, and erfc be the well-known complimentary error functions. Then R^α is absolutely continuous with respect to the Lebesgue measure on R_+, for $y \in R_+$

$$\mathbf{R}^\alpha(dy) = r^\alpha(y)dy,$$

where

$$r^\alpha(y) = \frac{\sigma}{\sqrt{y}} \varphi(\sqrt{y}\mu/\sigma) + (\frac{\mu - \alpha\sigma^2}{2}) e^{\alpha y(\frac{\alpha\sigma^2}{2} - \mu)} \operatorname{erf} c(\sqrt{y} \frac{\alpha\sigma^2 - \mu}{\sqrt{2\sigma^2}}). \quad (1.35)$$

Proof Let f be as defined in the proof of Lemma 1.18, then from (1.5) we have

$$\mathbf{R}^\alpha f(0) = \frac{\sigma^2}{\alpha\sigma^2 + \{\sqrt{2\theta\sigma^2 + \mu^2} - \mu\}}. \quad (1.36)$$

Our assertion is proven using Lemma 1.18 and inverting the right-hand side of (1.36) with respect to θ. ■

We now introduce the so-called *killed process*.

Definition 1.21 Let L be Lévy process and τ be a stopping time. For $t \geq 0$, let

$$X_t = \{L_t, t < \tau\}. \quad (1.37)$$

The process X is obtained by killing the process L at time τ.

Let X be the process defined in (1.37) then, for every Borel set A contained in the state space of X, $t \in R_+$, the probability transition function of this process is given as follows:

$$P_t(x, A) = P_x(L_t \in A, t < \tau\}$$

and for each $\alpha \in R_+$ its α-potential is defined as follows:

$$U^{\alpha}(x, A) = \int_0^{\infty} P_t(x, A)e^{-\alpha t}dt = E_x \int_0^{\tau} e^{-\alpha t}\mathbf{I}_{\{L_t \in A\}}dt. \tag{1.38}$$

For $\lambda \in R_+$, we define

$$T_{\lambda}^{+} = \inf\{t : L_t \geq \lambda\}. \tag{1.39}$$

If the stopping time τ in (1.37) is taken to T_{λ}^{+}, then the state space of the process X is $[0, \lambda)$ if it is a subordinate and $(-\infty, \lambda)$ if it is spectrally positive.

Lemma 1.22 Assume that the process L is a subordinator, and the process X is obtained by killing L at T_{λ}^{+}. For any Borel set $A \subset [0, \lambda)$, let $\mathbf{R}^{\alpha}(x, A)$ be as defined in (1.32), and $U^{\alpha}(x, A)$ be as defined in (1.38). Then, for $x \in [0, \lambda)$

$$U^{\alpha}(x, A) = \mathbf{R}^{\alpha}(x, A). \tag{1.40}$$

Proof Write

$$\begin{aligned} U^{\alpha}(x, A) &= E_x \int_0^{\infty} e^{-\alpha t}\mathbf{I}_{\{L_t \in A, t < T_{\lambda}^{+}\}}dt \\ &= E_x \int_0^{\infty} e^{-\alpha t}\mathbf{I}_{\{L_t \in A, , \bar{L}_t < \lambda\}}dt \\ &= E_x \int_0^{\infty} e^{-\alpha t}\mathbf{I}_{\{L_t \in A\}}dt \\ &= \mathbf{R}^{\alpha}(x, A), \end{aligned}$$

where the second equation above follows from the definition of T_{λ}^{+}, the third equation follows since, for each $t \geq 0$, $\bar{L}_t = L_t$ almost everywhere and $A \subset [0, \lambda)$. Furthermore, the last equation follows from (1.32). ■

As an application of the above result we have the following.

Theorem 1.23 Let X be a positive compound Poisson process as defined in Example 1 of Sect. 1.4. For $\alpha \geq 0$, let $\mathbf{R}^{\alpha}$ be as given in (1.34). For $x \geq 0$, let $\bar{\upsilon}(x) = \upsilon([x, \infty)) = \lambda \bar{F}(x)$, where $\bar{F} \equiv 1 - F$. Then for any $\lambda \geq 0$, $v \geq \lambda$, $u < \lambda$

$$E\{e^{-\alpha T_{\lambda}^{+}}, X_{T_{\lambda}^{+}} > v, X_{T_{\lambda}^{+}-} \leq u\} = \int_{(0,u]} \bar{\upsilon}(v - y)\mathbf{R}^{\alpha}(dy). \tag{1.41}$$

Proof For $n = 1, 2, ..,$ let $Y_n = X_1 + \cdots + X_n$, $Y_0 = 0$. Note that, for $n = 0, 1, ..,$ Y_n is the value of the compound Poison process at time S_n = time of the nth jump of the process X. Let N be the renewal process associated with $\{Y_n, n = 0, 1, .\}$, i.e. $N_x = \sup\{n : Y_n \leq x\}$. Furthermore, for $n = 1, 2, ...,$ S_n is a gamma random variable with mean n/λ. We write

$$
\begin{aligned}
E\{e^{-\alpha T_\lambda^+}, X_{T_\lambda^+} > v, X_{T_\lambda^+ -} \le u\}
&= \sum_{k=0}^{\infty} E[e^{-\alpha S_{k+1}}, Y_{k+1} > v, Y_k \le u, N_\lambda = k\} \\
&= \sum_{k=0}^{\infty} E[e^{-\alpha S_{k+1}}, Y_{k+1} > v, Y_k \le u\} \\
&= \sum_{k=0}^{\infty} E[e^{-\alpha S_{k+1}}, Y_k + X_{k+1} > v, Y_k \le u] \\
&= \sum_{k=0}^{\infty} E[e^{-\alpha S_{k+1}}, X_{k+1} > v - Y_k, Y_k \le u] \\
&= \sum_{n=0}^{\infty} E[e^{-\alpha S_{k+1}}] \int_{[0,u]} P\{X_{k+1} > v - y\} P\{Y_k \in dy\} \\
&= \sum_{k=0}^{\infty} (\frac{\lambda}{\lambda+\alpha})^{k+1} \int_{[0,u]} \bar{F}(v-y) P\{Y_k \in dy\} \\
&= \frac{\lambda}{\lambda+\alpha} \int_{[0,v]} \bar{F}(v-y) \sum_{nk=0}^{\infty} F_{\alpha}^{(k)}(dy) \\
&= \lambda \int_{[0,v]} \bar{F}(v-y) \mathbf{R}^{\alpha}(dy) \\
&= \int_{[0,u]} \bar{v}(v-y) \mathbf{R}^{\alpha}(dy).
\end{aligned}
$$

where the second equation follows since for every $k = 0, 1, ..., v > \lambda, u \le \lambda$, $\{Y_{k+1} > v, Y_k \le u\} \subset \{N_\lambda = k\}$, and the fifth equation follows since for $k = 0, 1, ..,$ the random variable S_{k+1} is independent of X_{k+1} and Y_k. ■

We conclude this section by computing the potential for spectrally positive processes. We start by computing the potential of a spectrally positive process killed at time T_λ^+. First, we let X be a spectrally positive Lévy process, and as usual we define $\hat{X} = -X$. For any $a \in R$, we let $T_a^- = \inf\{t \ge 0 : X_t \le a\}$, $\top_a^+ = \inf\{t \ge 0 : \hat{X}_t \ge a\}$, and $\top_a^- = \inf\{t \ge 0 : \hat{X}_t \le a\}$.

Lemma 1.24 Let X be a spectrally positive process, with α-scale function $W^{(\alpha)}$. For $\alpha \ge 0, a \le \lambda$ the α-potential $(\overset{(1)}{U^\alpha})$ of the process X killed at time $T = T_\lambda^+ \wedge T_a^-$ is absolutely continuous with respect to the Lebesgue measure on (a, λ) and a version of its density is given by

$$\overset{(1)}{u^{\alpha}}(x,y) = W^{(\alpha)}(\lambda - x)\frac{W^{(\alpha)}(y-a)}{W^{(\alpha)}(\lambda - a)} - W^{(\alpha)}(y-x), \quad x, y \in (a, \lambda). \quad (1.42)$$

Proof For any Borel set $A \subset (a, \lambda)$

$$\begin{aligned}
\overset{(1)}{U^{\alpha}}(x, A) &= E_x \int_0^{T_\lambda^+ \wedge T_a^-} e^{-\alpha t} \mathbf{I}_{\{I_t \in A\}} dt \\
&= E_{-x} \int_0^{T_{-\lambda}^- \wedge T_{-a}^+} e^{-\alpha t} \mathbf{I}_{\{\hat{I}_t \in -A\}} dt \\
&= E_{\lambda - x} \int_0^{T_0^- \wedge T_{\lambda - a}^+} e^{-\alpha t} \mathbf{I}_{\{\hat{I}_t \in \lambda - A\}} dt \\
&= \int_{(\lambda - A)} [W^{(\alpha)}(\lambda - x)\frac{W^{(\alpha)}(\lambda - a - y)}{W^{(\alpha)}(\lambda - a)} - W^{(\alpha)}(y - x)] dy,
\end{aligned}$$

where the last equation follows from Theorem 8.7 of [3], this establishes our assertion. ■

Corollary 1.25 Let X be a spectrally positive Lévy process, with α-scale function $W^{(\alpha)}$. For $\alpha \geq 0$ the α-potential (U^α) of the process killed at time T_λ^+ is absolutely continuous with respect to the Lebesgue measure on $(-\infty, \lambda)$ and a version of its density is given by

$$u^{\alpha}(x, y) = W^{\alpha}(\lambda - x)e^{-(\lambda - y)\eta(\alpha)} - W^{\alpha}(y - x), x, y \in (-\infty, \lambda). \quad (1.43)$$

Proof The proof follows from (1.42) by letting $a \to -\infty$ and since from (1.18), for $\alpha \geq 0$, $W^{(\alpha)}(x) \sim \frac{e^{\eta(\alpha)\,x}}{\phi'(\eta(\alpha))}$ as $x \to \infty$. ■

Corollary 1.26 Let X be a spectrally positive Lévy process, with α-scale function $W^{(\alpha)}$. For $\alpha \geq 0$ the α-potential $(\mathbf{R}^\alpha)$ is absolutely continuous with respect to the Lebesgue measure on $(-\infty, \infty)$ and a version of its density is given by

$$r^{\alpha}(x, y) = \frac{e^{-(x-y)\eta(\alpha)}}{\phi'(\eta(\alpha))} - W^{\alpha}(y - x), x, y \in (-\infty, \infty). \quad (1.44)$$

Proof The proof follows from (1.43) by letting $\lambda \to \infty$ and since, for $\alpha \geq 0$, $W^{(\alpha)}(\lambda - x) \sim \frac{e^{\eta(\alpha)(\lambda - x)}}{\phi'(\eta(\alpha))}$ as $\lambda \to \infty$. ■

The following is well known (see (8.8) of [3]), whose proof is outside the scope of this book and is omitted.

Lemma 1.27 Let $\hat{X}$ be a spectrally negative Lévy process, T_a^- and T_a^+ be the times of first hitting level a from above and below, respectively. Then, for $x \leq a$ and $\alpha \in R_+$,

$$E_x[e^{-\alpha \mathsf{T}_a^+}, \mathsf{T}_0^- > \mathsf{T}_a^+] = \frac{W^{(\alpha)}(x)}{W^{(\alpha)}(a)}. \tag{1.45}$$

References

1. Takács L (1968) On dams with finite capacity. J Aust Math Soc 8:161–170
2. Bertoin J (1996) Lévy processes. Cambridge University Press, Cambridge
3. Kyprianou AE (2006) Introductory lectures on fluctuations of Lévy processes and their applications. Springer, Berlin
4. Jacod J (1975) Multivariate point processes: predictable projection, Randon-Nikodym derivatives, representation of martingales. Z Wahrscheinlichkeitstheorie Verw Gebiete 31:235–253
5. Jacod J, Shiryaev A (2003) Limit theorems for stochastic processes. Springer, Berlin

Further Reading

6. Bernyk V, Danlang RC, Beskir G (2008) The law of the supremum of a stable Lévy process with no negative jump. Ann Probab 36:1777–1789
7. Feller W (1971) An Introduction to Probability Theory and its Applications, vol 2. Wiley, New York
8. Takács L (1967) Combinatorial methods in the theory of stochastic processes. Wiley, New York

Chapter 2
Degradation Processes

Abstract A device is subject to degradation. Over time, the degradation process is a nonhomogeneous subordinator. The device has a threshold (nominal life), and it fails once the degradation exceeds the threshold. We examine life distribution and failure rate properties of such devices, and determine optimal maintenance and replacement policies for such devices using the total and long-run average cost criteria. Inference about the parameters of the degradation process is also discussed.

Keywords Degradation processes · Nonhomogeneous subordinator · Semi-Markov processes · Semi-Markov kernel · Renewal function · Life distributions · Increasing failure rate · Increasing failure rate average · Maintenance policies · Maximum likelihood estimators · Moment estimators

2.1 Introduction

In reliability studies, the question of assessing the behavior of the failure rate of devices always arises. In practice, it is assumed that the life length of a given device has a certain distribution, Weibull, gamma, exponential, etc., or that it belongs to a certain class of distributions, such as increasing failure rate, increasing failure rate average, etc. Based on data collected about the failure times of identical devices, optimal estimates of the failure rate are obtained and hypothesis testing for the parameters of the assumed distribution is carried out. In many cases, collecting enough data to carry out sound statistical conclusions about the behavior of the failure rate and other parameters of the failure time distribution is not always possible. Even if enough data was available, the validity of the inference procedure is questionable because of the assumptions that are imposed on the distribution function. One way to avoid such difficulties is to examine and collect data on the failure mechanism of the given device, and make inference based on this data. Since many devices fail because of degradation, inference about the failure rate can be made based on the properties of the underlying degradation process.

In [1] the author proposed the nonhomogeneous gamma process as a model for describing degradation over time. He discussed life distributions of devices subject to this type of deterioration. Such degradation process received tremendous attention by reliability theorists and practitioners. The references at the end of this chapter contain a list of the relevant publications that appeared in the literature in this area over the last two decades. In [2] he discusses maintenance problems for devices subject to gamma deterioration process, where the deterioration is observed continuously as well as in discrete time intervals. In [3] the gamma degradation process is applied to maintenance of a cylinder on a bridge. In [4] periodic maintenance of systems subject to deterioration, where failure is detected only by inspection, using the long-run average cost criteria, is considered. Extensions of this model, assuming fixed failure level, are discussed in [5–8]. Similar maintenance models are discussed in [9–12].

The gamma process is found to model corrosion, crack growth, erosion, as well as creep. It has been used in [13] to model creep of concrete. It is also used in [14] to fitted data on fatigue crack control. It is used in [15] to describe corroded steel gates. In [16] the gamma process is used to determine optimal disk heightening. In [17] it is used to model steel pressure vessels. It is also used to model deterioration in automobile brake pads, as well as to determine optimal maintenance for steel coating.

An advantage of modeling deterioration processes through gamma processes is that the required mathematical calculations are relatively straightforward. One would hope that other degradation processes and maintenance policies of devices subject to such degradation processes will be explored by safety and maintenance practitioners. For example, the inverse Gaussian process, or compound Poisson process can be used to model degradation, the results obtained for the gamma degradation process can be easily extended to these case. As discussed in Sect. 1.4, these processes are special cases of Lévy subordinators. In general, degradation processes can be best described using Lévy subordinators.

In this chapter, we examine, in details, how nonhomogeneous subordinators can be used to describe degradation over time. In Sect. 2.3 we examine life distribution properties of devices subject to such degradation processes. Specifically, we examine the behavior of the failure rates of devices subject to a nonhomogeneous subordinator deterioration, and give conditions that insure that the failure time belongs to the different classes of life distributions. In Sect. 2.4, we examine maintenance and replacement policies of devices subject to such degradation using the total discounted as well as the long-run average cost criteria. In Sect. 2.5, some special cases are discussed. In Sect. 2.6, inference about the parameters of some degradation processes and other related results, are discussed.

2.2 Basic Definitions and Results

In this section, we give some basic reliability definitions, and discuss some well-known results that will be used in the rest of this chapter. Detailed proofs of these results are found in the references indicated at the end on this chapter.

Let X be a positive random variable describing the life time of a given device. For any $x \geq 0$, the survival probability (reliability), by definition, $\bar{F}(x) = P\{X > x\}$. Suppose that $A = \{x \in R_+ : \bar{F}(x) > 0\}$. We define $\frac{0}{0} = 0$, throughout the rest of this chapter.

Definition 2.1 A survival probability $\bar{F}$ is said to be or to have:

(i) *Increasing Failure Rate* (IFR) if, for each $\triangle \geq 0$, $\frac{\bar{F}(t+\triangle)}{\bar{F}(t)}$ is decreasing in $t \in A$. Equivalently, the hazard function $R \overset{def}{=} -ln\bar{F}$ is convex. When $\bar{F}$ has a density, this is equivalent to the condition that, for some form of the density f the failure rate $r(t) \overset{def}{=} \frac{f(t)}{\bar{F}(t)}$ is increasing in $t \geq 0$. To say that the survival probability of a device is IFR is equivalent to saying that the residual life length of an unfailed device of age t is stochastically decreasing in $t \geq 0$.

(ii) *Increasing Failure Rate Average* (IFRA) if, $[\bar{F}(t)]^{1/t}$ is decreasing in $t \in A$. Equivalently, the hazard function R is starshaped on $[0, \infty)$. (A positive function $g : R_+ \to R_+$with $g(0) = 0$, is said to be starshaped if, for $\alpha \in [0, 1]$, and $x \in R_+$, $g(\alpha x) \leq \alpha g(x)$). When the failure rate exists, this is equivalent to saying that the average failure rate $\frac{1}{t}\int_0^t r(u)du$ is increasing in $t \geq 0$.

(iii) *New Better than Used* (NBU) if, for all $t, x \in R_+$, $\bar{F}(x) \geq \frac{\bar{F}(t+x)}{\bar{F}(t)}$. Equivalently, the hazard function R is superadditive (A positive function $g : R_+ \to R_+$ with $g(0) = 0$, is said to be superadditive if, for all $x, y \in R_+$, $g(x + y) \geq g(x) + g(y)$). To say that a survival probability is NBU is equivalent to saying that the life length of a new device is stochastically greater than that of unfailed device of age t.

We note that if a positive function f is convex and $f(0) = 0$, then f is necessarily starshaped. Furthermore, if f is starshaped, then f must be superadditive. The reverse implication does not hold as seen by choosing $f(x) = [x]$ a superadditive function which is not starshaped. Hence, we have the following implications

$$IFR \Longrightarrow IRFA \Longrightarrow NBU. \tag{2.1}$$

There are dual life distribution classes parallel to the above-mentioned classes. They are obtained by reversing the direction of inequality or monotonicity in the above definition. These classes are the decreasing failure rate (DFR), decreasing failure rate average (DFRA), and new worse than used (NWU) classes. Note that $\bar{F}$ is DFRA if and only if hazard function R is anti-starshaped on $[0, \infty)$. (A positive function $g : R_+ \to R_+$ with $g(0) = 0$, is said to be anti-starshaped if, for $\alpha \in [0, 1]$, and $x \in R_+$, $g(\alpha x) \geq \alpha g(x)$). Furthermore, $\bar{F}$ is NWU if and only if the hazard function R is subadditive (A positive function $g : R_+ \rightarrow R_+$ with $g(0) = 0$, is said to be subadditive if, for all $x, y \in R_+$, $g(x + y) \leq g(x) + g(y)$).

Knowing the behavior of the failure rate of any device enables us to determine the appropriate maintenance and replacement policies for such device.

Examples of IFR random variables are: The Weibull random variable with shape parameter $\alpha \geq 1$, the gamma random variable with shape parameter $\alpha \geq 1$, the truncated normal random variable with support R_+.

Life distributions of coherent systems of independent components having respective exponential life times is IFRA.

DFR random variables include: mixtures of exponential random variables, the Weibull random variable with shape parameter $\alpha \leq 1$, and the gamma random variable with shape parameter $\alpha \leq 1$.

A positive random variable, with survival function $\bar{F}(t) = \exp(-a[bt]), a, b \geq 0$ is NBU, where $[x]$ is the greatest integer greater than or equal to x.

Dependence of Random variables. Components of systems exhibit some degree of dependence between their performances, indicated by their life times. This dependence could be positive, as in the failure times of components subject to the same environment. Negative dependence arises in competing risk applications, where items are competing for a fixed amount of resources. The statistical literature is full of different measures of dependence. The simplest is the correlation and partial correlation coefficients. The notion of association between random variables (Sect. 2.2 of [18]) has many applications in reliability and statistics. While there are many other measures of dependence between random variables, this is not the place to go thoroughly through them. One of the strongest notions of positive dependence is given in the following

Definition 2.2 A function $f : R^2 \longrightarrow R$, is said to be totally positive of order 2 (TP_2) if $det(f(x_i, y_j)) \geq 0$ for each choice $x_i \leq x_2$, and $y_i \leq y_2$.

The proof of the following follows easily from Definition 2.2 and is omitted.

Corollary 2.3 Let $f : R^2 \longrightarrow R$, assume that f is differentiable with respect to both of its arguments, then f is TP_2 if and only if $\frac{\partial^2}{\partial x \partial t} \ln f(t, x) \geq 0$.

Example 1 Let X be a gamma process given in Example 3 of Sect. 1.4, then its transition function is of the form

$$p(t, x) = \beta e^{-x\beta} \frac{(x\beta)^{t-1}}{\Gamma(t)}.$$

It follows that $\frac{\partial^2}{\partial x \partial t} \ln p(t, x) = \frac{1}{x} \geq 0$, hence $p(t, x)$ is TP_2.

Example 2 Let X be an inverse Gaussian process given in Example 2 of Sect. 1.4, its transition function is of the form

$$p(t,x)=\begin{cases}\frac{t}{\sigma\sqrt{2\pi(x)^3}}\exp\{-\frac{[\mu x-t]^2}{2x\sigma^2}\}, & x>0\\ 0, & x\le 0.\end{cases}$$

It follows that $\frac{\partial^2}{\partial x\partial t}\ln p(t,x)=\frac{t}{(x\sigma)^2}\ge 0$. Thus, $p(t,x)$ is TP_2.

Example 3 Let X be a positive compound Poisson process, and assume that the jump distribution density f is PF_2 (in the sense described in Definition 2.4 below), then the transition function of X is TP_2. First we note that the total positivity of the transition density function is invariant under change of the arrival rate λ, so we can assume without loss of generality that $\lambda=1$. We write

$$p(t,x)=\sum_{k=0}^{\infty}e^{-t}\frac{t^k}{k!}f^{(k)}(x).$$

To show that $p(t,x)$ is TP_2, we proceed as follows: Note that for $t_1\le t_2$, and $x_1\le x_2$,

$$det(p(t_i,x_j))=\begin{vmatrix}\sum_{k=0}^{\infty}\frac{t_1^k}{k!}f^{(k)}(x_1) & \sum_{k=0}^{\infty}\frac{t_1^k}{k!}f^{(k)}(x_1)\\ \sum_{k=0}^{\infty}\frac{t_2^k}{k!}f^{(k)}(x) & \sum_{k=0}^{\infty}\frac{t_2^k}{k!}f^{(k)}(x_2)\end{vmatrix}$$

$$=\sum\sum_{0\le k_1<k_2<\infty}\begin{vmatrix}\frac{t_1^{k_1}}{k_1!} & \frac{t_1^{k_2}}{k_2!}\\ \frac{t_2^{k_1}}{k_1!} & \frac{t_2^{k_2}}{k_2!}\end{vmatrix}\begin{vmatrix}f^{(k_1)}(x_1) & f^{(k_2)}(x_1)\\ f^{(k_1)}(x_2) & f^{(k_2)}(x_2)\end{vmatrix}$$

where the last equation follows from Theorem 2.5 below. For $t_1\le t_2$ and $k_1\le k_2$, the first determinant is positive, and the second determinant is positive, since $f(x)$ is PF_2 in x, implies that $f^{(k)}(x)$ is TP_2 in x,k. It follows that for $t_1\le t_2$, and $x_1\le x_2$, $det(p(t_i,x_j))\ge 0$. Hence p is TP_2.

If f is the probability density function of two random variables, then f is TP_2, implies that the two random variables, loosely speaking, exhibit a very strong positive dependence. A more detailed investigation of this matter can be found in Sect. 5.4 of [18].

Definition 2.4 A function $f:R\longrightarrow R$ is said to be a Polya frequency function of order 2 (PF_2) if $f(x-y)$ is TP_2 in x,y.

Let $f,g:R^2\to R$, be two Borel measurable functions. Consider the transformation

$$h(x,y)=\int f(x,z)g(z,y)\mu(dz),$$

where μ is the Lebesgue measure on R or a counting measure, and the integral is assumed to exist. When μ is a counting measure, the integral is interpreted as a sum.

The following theorem is a special case of the result stated as (2.5) of [19] *as the Basic Composition Formula.*

Theorem 2.5 Let $h : R^2 \to R$, be as defined in the integral equation above. Then, for any $n \geq 2$, and $x_1, x_2, y_1, y_2 \in R$

$$det(h(x_i, y_j)) = \iint_{z_1<z_2} det(f(x_i, z_k))det(f(z_k, y_j))\mu(dz_1)\mu(dz_2).$$

Proof For $x_1, x_2, y_1, y_2 \in R$, we have

$$\begin{aligned} h(x_1, y_1)h(x_2, y_2) &= \iint_{R\times R} f(x_1, z_1)f(x_2, z_2)g(z_1, y_1)g(z_2, y_2)\mu(dz_1)\mu(dz_2), \\ h(x_1, y_2)h(x_2, y_1) &= \iint_{R\times R} f(x_1, z_1)f(x_2, z_2)g(z_1, y_2)g(z_2, y_1)\mu(dz_1)\mu(dz_2). \end{aligned} \tag{2.2}$$

For $u_1, u_2, y_1, y_2 \in R$, we define

$$g\begin{pmatrix} u_1 & u_2 \\ y_1 & y_2 \end{pmatrix} = \begin{vmatrix} g(u_1, y_1) & g(u_1, y_2) \\ g(u_2, y_1) & g(u_2, y_2) \end{vmatrix}$$

From (2.2), it follows that

$$\begin{aligned} det(h(x_i, y_j)) &= \iint_{R\times R} f(x_1, z_1)f(x_2, z_2)g\begin{pmatrix} z_1 & z_2 \\ y_1 & y_2 \end{pmatrix}\mu(dz_1)\mu(dz_2) \\ &= \iint_{z_1\leq z_2} f(x_1, z_1)f(x_2, z_2)g\begin{pmatrix} z_1 & z_2 \\ y_1 & y_2 \end{pmatrix}\mu(dz_1)\mu(dz_2) \\ &\quad + \iint_{z_2\leq z_1} f(x_1, z_1)f(x_2, z_2)g\begin{pmatrix} z_1 & z_2 \\ y_1 & y_2 \end{pmatrix}\mu(dz_1)\mu(dz_2) \\ &= \iint_{z_1\leq z_2} f(x_1, z_1)f(x_2, z_2)g\begin{pmatrix} z_1 & z_2 \\ y_1 & y_2 \end{pmatrix}\mu(dz_1)\mu(dz_2) \\ &\quad + \iint_{z_1\leq z_2} f(x_1, z_2)f(x_2, z_1)g\begin{pmatrix} z_2 & z_1 \\ y_1 & y_2 \end{pmatrix}\mu(dz_2)\mu(dz_1) \\ &= \iint_{z_1\leq z_2} det(f(x_i, z_k))det(f(z_k, y_j))\mu(dz_1)\mu(dz_2), \end{aligned}$$

where the last equation follows from the third equation, since $g\begin{pmatrix} z_2 & z_1 \\ y_1 & y_2 \end{pmatrix} = -g\begin{pmatrix} z_1 & z_2 \\ y_1 & y_2 \end{pmatrix}$. ■

Coherent Structures and System Reliability: Consider a system of n components. At any time instant any component can be either functioning or faulty. For each $i \in \{1, 2, ...n\}$ we let

$$X_i = \begin{cases} 0, & \text{if component i is faulty;} \\ 1, & \text{if component i is functioning,} \end{cases}$$

and $p_i = P\{X_i = 1\}$, and $q_i = 1 - p_i = P\{X_i = 0\}$.

Let $\underline{X} = \{X_1, ...X_n\}$. At any instant the system is either functioning or faulty. The *structure function* of the system $\phi : \{0, 1\}^n \to \{0, 1\}$ is defined as follows:

$$\phi(\underline{x}) = \begin{cases} 0, & \text{if system is faulty;} \\ 1, & \text{if system is functioning.} \end{cases}$$

The simplest systems are *series* and parallel systems. For series system, we have

$$\begin{aligned} \phi(\underline{x}) &= \min_{1 \le i \le n} x_i \\ &= \prod_i x_i . \end{aligned}$$

For parallel systems,

$$\begin{aligned} \phi(\underline{x}) &= \max_{1 \le i \le n} x_i \\ &= 1 - \prod_i (1 - x_i). \end{aligned}$$

Definition 2.6 A system is said to be coherent if

(i) $\phi(\underline{x})$ is increasing in $\underline{x}$, i.e. for any two vectors of binary components $\underline{x}_1 = (x_{11}, x_{12}, ..., x_{1n}\}$, and $\underline{x}_2 = (x_{21}, x_{22}, ..., x_{2n}\}$ such that $x_{1i} \ge x_{2i}$, $1 \le i \le n$, $\phi(\underline{x}_1) \ge \phi(\underline{x}_2)$,

(ii) each component is relevant, i.e. for each $i \in \{1, 2, ...n\}$ there exists an $\underline{x}$ such that

$$\phi(1_i.\underline{x}) > \phi(0_i.\underline{x}),$$

where $(1_i.\underline{x}) = (x_1, \ldots, x_{i-1}, 1, x_{i+1}, \ldots, x_n)$, and $(0_i.\underline{x}) = (x_1, \ldots, x_{i-1}, 0, x_{i+1}, \ldots, x_n)$.

If a component in a system of n components is irrelevant, then removing it from the system would result in a simpler system of order less than or equal to $n-1$ that has identical performance.

The reliability of the system is equal to $E[\phi(\underline{X})]$.

In the above discussion, we dealt with a system whose reliability is the same over time. In practice, this not this case and the reliability of the system is time-dependent. Assume that a system has n components with structure function ϕ. For $i \in \{1, 2, ...n\}$, we let T_i be the positive random variable describing the failure time of the ith component; for each $t \geq 0$, we define the random variable $X_i(t)$ as follows:

$$X_i(t) = \mathbf{I}_{\{T_i > t\}}. \tag{2.3}$$

We also define, $\tau(T_1, ..., T_n\}$ as the failure time of the system. For $t \geq 0$, the survival probability of the system, $\bar{F}\ (t)$, is given as follows:

$$\begin{aligned}\bar{F}(t) \stackrel{def}{=} P\{\tau(T_1, ..., T_n\} > t\} &= E[\phi(\underline{X}(t))] \\ &= E[\phi(\mathbf{I}_{\{T_1>t\}}, ..., \mathbf{I}_{\{T_n>t\}})]. \end{aligned} \tag{2.4}$$

The following theorem is a restatement of Theorem 2.13 of [18].

Theorem 2.7 A survival probability $\bar{F}$ is IFRA, if and only $\bar{F}(t) = \lim_{n\to\infty} \bar{F}_n(t)$, where, for each $n \geq 1$, $\bar{F}_n$ is the survival probability of a coherent system of n components, and the distribution of the failure time of each component is exponential.

2.3 Life Distributions of Devices Subject to Degradation

Assume that over time, a device is subject to degradation. For each $t \geq 0$, X_t is the amount of degradation the device suffers during $[0, t]$. The degradation process $X = \{X_t, t \in R_+\}$ is a nonhomogeneous increasing stochastic process. In this case, for each $t \geq 0$,

$$X_t = \hat{X}_{\Lambda(t)}, \tag{2.5}$$

where $\Lambda : R_+ \to R_+$ is increasing, and the process $\hat{X}$ is a homogeneous increasing stochastic process. The device has a threshold Y, and it fails once the degradation exceed or equal Y. For each $x \in R_+$, we define $\bar{G}(x) = P\{Y > x\}$. Let ς be the random variable denoting the life time of the device. Note that $\varsigma = \inf\{t \geq 0 : X_t \geq Y\}$. For each $t \in R_+$, the survival probability $\bar{F}(t)$ is given as follows:

$$\begin{aligned}\bar{F}(t) &= P\{\varsigma > t\} = P\{Y > X_t\} \\ &= E[\bar{G}(X_t)]. \end{aligned} \tag{2.6}$$

For each $t \in R_+$, we let

$$\begin{aligned}\bar{H}(t) &= P\{Y > \hat{X}_t\} \\ &= E[\bar{G}(\hat{X}_t)]. \end{aligned} \tag{2.7}$$

Note that, for each $t \in R_+$,

$$\bar{F}(t) = \bar{H}(\Lambda(t)).$$

The following theorem gives necessary and sufficient conditions that guarantees that the degradation process is a subordinator and sufficient conditions for the life distribution of devices subject to nonhomogeneous subordinator degradation process to be Weibull.

Theorem 2.8 (i) The degradation process is a subordinator, if and only if $\bar{G}$ exponential implies that $\bar{H}$ is exponential.
(ii) Assume that the degradation process is a nonhomogeneous subordinator. Then, $\bar{F}$ is Weibull if $\bar{G}$ is exponential and $\Lambda(t) = t^{\beta}$, $\beta > 0$.

Proof (i) $\Longrightarrow$ Assume that the degradation process $\hat{X}$ is a stationary subordinator, and $\bar{G}$ exponential. From (1.3), it follows that $\bar{H}$ is exponential.
$\Longleftarrow$ Assume that $\bar{G}$ exponential implies that $\bar{H}$ is exponential. Suppose $\bar{G}(x) = e^{-\lambda x}$, $\lambda, x \geq 0$. If $\bar{H}$ is exponential then, for each $t, s \geq 0$, $\bar{H}(t+s) = \bar{H}(t)\bar{H}(s)$. For $t \geq 0$, $A \subset R_+$, let $P_t(A) = P(\hat{X}_t \in A)$. But, $\bar{H}(t)\bar{H}(s) = \int_{[0,\infty)} e^{-\lambda x} P_t * P_s(dx)$. From the uniqueness of the Laplace-Stieltjes transform and since $\bar{H}(t+s) = \int_{[0,\infty)} e^{-\lambda x} P_{t+s}(dx)$, it follows that for $t, s \geq 0$ and $A \subset R_+$, $P_{t+s}(A) = P_t * P_s(A)$, thus $\hat{X}$ is a stationary subordinator.

(ii) The proof is immediate from (1.30). ■

Theorem 2.9 Let $\bar{F}$ be as given in (2.6). Assume that the degradation process is a nonhomogeneous subordinator. Then,

(i) If $\bar{G}$ is IFR, for each $t \in R_+$, the distribution function $P(t, x) = P\{\hat{X}_t \leq x\}$ is absolutely continuous, in x, with respect to the Lebesgue measure on R_+ and version of its density is TP_2, and $\Lambda(t)$ is convex, then $\bar{F}$ is IFR.
(ii) If $\bar{G}$ is IFRA, and $\Lambda(t)$ is starshaped, then $\bar{F}$ is IFRA.
(iii) If $\bar{G}$ is NBU, and $\Lambda(t)$ is superadditive, then $\bar{F}$ is NBU.

Proof (i) Let $\hat{R} = -\ln \bar{F}$, $R = -\ln \bar{H}$, and note that for each $t \geq 0$, $\hat{R}(t) = R(\Lambda(t))$. Since convex increasing function of a convex function is convex, it is sufficient to prove that R is convex, i.e., $\bar{H}$ is IFR. For any positive t_1,t_2, and Δ we define

$$D(\Delta, t_1,\ t_2) = \bar{H}(t_1 + \Delta)\bar{H}(t_2) - \bar{H}(t_2 + \Delta)\bar{H}(t_1).$$

We note that $\bar{H}$ is IFR if and only if $D(\Delta, t_1,\ t_2) \geq 0$, for each $t_1 \leq t_2$. Let $p(t, x)$ be the probability transition function of the process $\hat{X}$. For any positive t, x we let

$$f(t, x) = \int_0^\infty p(t, z)\bar{G}(x + z)dz$$

For any positive t, Δ, we have

$$\bar{H}(t + \Delta) = \int_0^\infty p(t + \Delta, x)\bar{G}(x)dx$$
$$= \int_0^\infty p(\Delta, x) f(t, x)dx,$$

where the last equality follows since $\hat{X}$ has stationary independent increments. From the last equality, it follows $D(\Delta, t_1,\ t_2) \geq 0$ if the determinant

$$f\begin{pmatrix} t_1 & t_2 \\ x & 0 \end{pmatrix} = \begin{vmatrix} f(t_1, x) & f(t_x, x) \\ f(t_1, 0) & f(t_2, 0) \end{vmatrix}$$

is positive. From Theorem 2.5 we have

$$f\begin{pmatrix} t_1 & t_2 \\ x & 0 \end{pmatrix} = \iint\limits_{z_1 \leq z_2} \begin{vmatrix} p(t_1, z_1) & p(t_1, z_2) \\ p(t_2, z_1) & p(t_2, z_2) \end{vmatrix} \begin{vmatrix} \bar{G}(x + z_1) & \bar{G}(x + z_2) \\ \bar{G}(z_1) & \bar{G}(z_2) \end{vmatrix}.$$

Our assertion is proven since, for $t_1 \leq t_2$, the first determinant (under the integral sign) is positive, because $p(t, x)$ is TP_2; furthermore the second determinant is also positive, since $\bar{G}$ is IFR. ■

(ii) From the definition of the functions R and $\hat{R}$ above, we have: for each $t \geq 0$, and $\alpha \in (0, 1)$, $\hat{R}(\alpha t) = R(\Lambda(\alpha t))$. Proving that $\bar{F}$ is IFRA is equivalent to proving that $\hat{R}$ is starshaped. Since, Λ is starshaped and R is increasing, then for each $t \geq 0$ and $\alpha \in (0, 1)$, $R(\Lambda(\alpha t)) \leq R(\alpha\Lambda(t))$. Thus, it suffices to prove that R is starshaped, i.e., $\bar{H}$ is IFRA. Since $\bar{G}$ is IFRA, from Theorem 2.7 it follows that , for each $x \in R_+$,

$$\bar{G}(x) = \lim_{n\to\infty} P\{\tau(Y_1, ..., Y_n\}>x\}$$

$$= \lim_{n\to\infty} E[\phi(\mathbf{I}_{\{Y_1>x\}}, ..., \mathbf{I}_{\{Y_n>x\}}].$$

where $\tau(Y_1, ..., Y_n$ is the life time of a coherent system of n components, with exponential failure times $Y_1, ..., Y_n$, with structure function ϕ.

For $1 \leq i \leq n$, let

$$T_i = \inf\{t : \hat{X}_t > Y_i\}.$$

Since Y_i is an exponential random variable, then for some θ_i, $P\{Y_i > y\} = e^{-\theta_i\, y}$, $y \geq 0$. Thus,

$$\begin{aligned} P\{T_i > t\} &= P\{Y_i > \hat{X}_t\} \\ &= E[e^{-\theta_i \hat{X}_t}] \\ &= e^{-t\psi(\theta_i)}, \end{aligned}$$

where ψ is the Laplace exponent of the process $\hat{X}_t$. Hence T_i is an exponential random variable.
From (2.7) we have

$$\begin{aligned} \bar{H}(t) &= E[\bar{G}(\hat{X}_t)] \\ &= \lim_{n\to\infty} E[\phi(\mathbf{I}_{\{Y_1>\hat{X}_t\}}, ..., \mathbf{I}_{\{Y_n>\hat{X}_t\}}] \\ &= \lim_{n\to\infty} E[\phi(\mathbf{I}_{\{T_i>t\}}, ..., \mathbf{I}_{\{T_n>t\}}]. \end{aligned}$$

Thus, $\bar{H}$ is the limit of survival probabilities of coherent systems whose components have exponential failure times. Our assertion that $\bar{H}$ is IFRA follows from Theorem 2.7.

(iii) Since *Lambda* is superadditive, and R is increasing, for $s, t \in R_+$, we have

$$\begin{aligned} \hat{R}(t+s) &= R(\Lambda(t+s)) \\ &\leq R(\Lambda(t) + \Lambda(s)). \end{aligned}$$

Since $\bar{F}$ is NBU if and only if $\hat{R}$ is superadditive, then it suffices to show that R is superadditive, i.e., $\bar{H}$ is NBU.

For $s, t \in R_+$, we have

$$\begin{aligned}
\bar{H}(t+s) &= E[\bar{G}(\hat{X}_{t t+s})] \\
&= E[\bar{G}(\hat{X}_{t+s} - \hat{X}_s + \hat{X}_s)] \\
&\leq E[\bar{G}(\hat{X}_{t+s} - \hat{X}_s)\bar{G}(\hat{X}_s)] \\
&= E[\bar{G}(\hat{X}_t)]E[\bar{G}(\hat{X}_s)] \\
&= \bar{H}(t)\bar{H}(s),
\end{aligned}$$

where the first equality follows from the definition of $\bar{H}$, the third inequality follows since $\bar{G}$ is NBU, and the fourth equality follows since the processes $\{\hat{X}_t, t \geq 0\}$ has stationary independent increments. ■

The following theorem provides results similar to those in Theorem 2.9 for the dual classes of life distributions. The proof of this theorem, which we omit follows by modifying the proof of Theorem 2.9 in obvious manners.

Theorem 2.10 Let $\bar{F}$ be as given in (2.6). Assume that the degradation process is a nonhomogeneous subordinator. Then,

(i) If $\bar{G}$ is DFR, for each $t \in R_+$, the distribution function $P(x,t) = P\{\hat{X}_t \leq x\}$ is absolutely continuous with respect to the Lebesgue measure on R_+, version of its density is TP_2, and $\Lambda(t)$ is concave, then $\bar{F}$ is DFR.
(ii) If $\bar{G}$ is DFRA, and $\Lambda(t)$ is anti-starshaped, then $\bar{F}$ is DFRA.
(iii) If $\bar{G}$ is NWU, and $\Lambda(t)$ is subadditive, then $\bar{F}$ is NWU.

2.4 Control-Limit Maintenance Policies for Continuously Monitored Degradable Systems

Assume that a system is subject to deterioration, the deterioration is continuously monitored. The deterioration process is a nonhomogeneous subordinator, denoted by X. As discussed in Sect. 2.3, for each $t \in R_+$, $X_t = \hat{X}_{\Lambda(t)}$, where $\hat{X}$ is a time homogeneous subordinator and $\Lambda : R_+ \to R_+$, is increasing. Let Y be the random variable describing the threshold (resistance) of the device, with right tail probability $\bar{G}$. Suppose that $B = \{x \in R_+ : \bar{G}(x) > 0\}$, and define

$$\zeta = \inf\{t \geq 0 : X_t \geq Y\} \text{ and } \hat{\zeta} = \inf\{t \geq 0 : \hat{X}_t \geq Y\}.$$

Let κ be the right-continuous inverse of Λ, i.e., $\kappa(t) = \inf\{s : \Lambda(s) > t\}$. It follows that $\zeta = \kappa(\hat{\zeta})$, $\hat{\zeta} = \Lambda(\zeta)$.

For $t \in R_+$, we let

$$Z_t = \begin{cases} X_t, \ t < \zeta, \\ \infty. \ t \geq \zeta, \end{cases} \quad \text{and} \quad \hat{Z}_t = \begin{cases} \hat{X}_t, \ t < \hat{\zeta}, \\ \infty. \ t \geq \hat{\zeta}, \end{cases}$$

$F_t = \sigma(Z_s, s \leq t)$ and $\hat{F}_t = \sigma(\hat{Z}_s, s \leq t)$. It follows that, for each $t \geq 0$, $Z_t = \hat{Z}_{\Lambda(t)}$. Furthermore, both $Z$and $\hat{Z}$ are strong Markov processes.

The following maintenance and replacement structure is adopted:

(1) A replacement time is defined to be any stopping time with respect to the history $F = \{F_t, t \geq 0\}$, that is less than or equal to ζ almost everywhere. That is to say, at any time $t \geq 0$, the decision to replace or not to replace depends only on the values of Z up to time$t$.
(2) The device can be replaced at or before a failure occurs.
(3) The cost of a replacement at failure is equal to a constant $c > 0$.
(4) The cost of a replacement before failure depends on the deterioration level at the time of replacement and is denoted by $c_1(x)$, $c_1(x) < c$ on B, and is equal to c on the complement of with respect to $\bar{R}_+$.

We will denote the complement of B with respect to $\bar{R}_+$ by $\bar{B}$, throughout this section.

Proposition 2.11 Let Ψ be the class of stopping times, with respect to the history F_t, that are less than or equal to ζ, and $\hat{\Psi}$ be the class of stopping times, with respect to $\hat{F}_t$, that are less than or equal to $\hat{\zeta}$. Then, a replacement time $\tau \in \Psi$ if and only if $\Lambda(\tau) \in \hat{\Psi}$.

Proof Note that, for each $t \geq 0$,

$$\begin{aligned} F_t &= \sigma(Z_s, s \leq t) \\ &= \sigma(\hat{Z}_{\Lambda(s)}, s \leq t) \\ &= \sigma(\hat{Z}_u, \kappa(u) \leq t) \\ &= \sigma(\hat{Z}_u, u \leq \Lambda(t)) \\ &= \hat{F}_{\Lambda(t)}. \end{aligned}$$

To prove the only if part, assume that $\tau \in \Psi$, then for each $t \geq 0$, $\{\tau > t\} \in \mathcal{F}_t = \hat{\mathcal{F}}_{\Lambda(t)}$. Denote, $\Lambda(\tau)$ by $\hat{\tau}$, then

$$\begin{aligned}\{\hat{\tau} > t\} &= \{\Lambda(\tau) > t\} \\ &= \{\tau > \kappa(t)\} \in \mathcal{F}_{\kappa(t)} = \hat{\mathcal{F}}_t.\end{aligned}$$

Furthermore, $\tau \leq \zeta$ if and only $\hat{\tau} \leq \hat{\zeta}$. This finishes the proof of the only if assertion. The proof of the if part follows in a similar way with obvious modifications. ■

Let $\tau \in \Psi$, and assume that, $E[\tau] < \infty$. Using standard renewal argument (see Theorem 3.6.1 of [20]), it follows that the long-run average cost, using a policy τ, denoted by $\varphi(\tau)$ is as follows:

$$\varphi(\tau) = \frac{cP\{\tau = \zeta\} + E[c_1(Z_\tau), \tau < \zeta]}{E[\tau]}.$$

We note that a sufficient condition for $E[\tau]$ to be finite is that $E[\zeta]$ is finite. The following theorem gives sufficient conditions that insure that $E[\zeta]$ is finite.

Theorem 2.12 Assume that $\bar{G}$ is IFRA, and Λ is starshaped. Then, $E[\zeta] < \infty$.

Proof From (ii) of Theorem 2.9, it follows that $\bar{F}$ is IFRA. Therefore, $-ln\bar{F}$ is starshaped. Hence, there exists a $\lambda, t^* \in R_0$, such that $\bar{F}(t) \geq e^{-\lambda t}$ for $t < t^*$, and $\bar{F}(t) \leq e^{-\lambda t}, t \geq t^*$. Thus,

$$\begin{aligned}E[\zeta] &= \int_0^\infty P\{\zeta > t\} \\ &= \int_0^\infty \bar{F}(t)dt \\ &\leq \int_0^{t^*} \bar{F}(t)dt + \int_{t^*}^\infty e^{-\lambda t}dt \\ &< \infty,\end{aligned}$$

and our assertion is now proven. ■

Definition 2.13 Let X be a Markov process. The infinitesimal generator of the process Y, denoted by A, is an operator such that for any bounded function f

$$Af(x) = \lim_{t \downarrow 0} \frac{E_x[f(X_t)] - f(x)}{t}.$$

The domain of the generator, denoted by $Ð_A$, is defined as the class of bounded functions f for which $Af(x)$ exists and is bounded.

The following is known as *Dynkin's Lemma.* (see Theorem 7.4.1 of [21]).

Theorem 2.14 Let X be a strong Markov process, with generator A. Then, for every stopping time τ, such that $E_x[\tau] < \infty$, and $f \in Ð_A$,

$$E_x[f(X_\tau)] - f(x) = E_x[\int_0^\tau Af(X_s)ds].$$

Proposition 2.15 Let A be the infinitesimal generator of the process $\hat{Z}_t$. Let $\mathcal{F} = \{f: R_+ \cup \{\infty\} \to R,\ f$ is bounded, and has bounded derivatives, and $f \equiv 0$ on $\bar{B}\}$. Assume that $\bar{G}$ is absolutely continuous on B, with bounded failure rate. If $f \in \mathcal{F}$, then $f \in Ð_A$ and for $x \in B$

$$\begin{aligned} Af(x) &= \frac{1}{\bar{G}(x)} \int_B [f(x+y)\bar{G}(x+y) - f(x)\bar{G}(x)]\upsilon(dy),\ x \in B, \\ &= 0,\ x \in \bar{B}. \end{aligned}$$

where ν is the Lévy measure of the process $\hat{X}$.

Proof The fact that $Af(x) = 0$ outside B follows, since $f \equiv 0$ on $\bar{B}$. Let $h = f \times \bar{G}$, since $f(\infty) = 0$, then for each $t \geq 0$, and $x \in B$, we have

$$\begin{aligned} E_x[f(\hat{Z}_t)] - f(x) &= E_x[f(\hat{Z}_t), t < \overset{\wedge}{\zeta}] - f(x) \\ &= \frac{1}{\bar{G}(x)}\{E_x[f(\hat{X}_t)\bar{G}(\hat{X}_t)] - f(x)\bar{G}(x)\} \\ &= \frac{1}{\bar{G}(x)}\{E_x[h(\hat{X}_t)] - h(x)\}. \end{aligned}$$

Let M be the Poisson random measure corresponding to the homogeneous subordinator $\hat{X}$. Write

$$E_x[h(\hat{X}_t)] - h(x) = E_x[\int_{[0,t]\times B} \{h(y + \hat{X}_{s-}) - h(\hat{X}_{s-})\}M(ds, dy)]$$

$$= E_x[\int_{[0,t]\times B} \{h(y+\hat{X}_s) - h(\hat{X}_s)\} ds \nu(dy)],$$

where the last equation follows from the Theorem 1.16. Thus, for $x \in B$,

$$\begin{aligned} Af(x) &= \lim_{t\downarrow 0} \frac{E_x[f(\hat{Z}_t)] - f(x)}{t} \\ &= \bar{G}^{-1}(x) \lim_{t\downarrow 0} \{E_x[h(\hat{X}_t)] - h(x)\} \\ &= \bar{G}^{-1}(x) \int_{R_0} [h(x+y) - h(x)]\nu(dy) \\ &= \bar{G}^{-1}(x) \int_{B} [f(x+y)\bar{G}(x+y) - f(x)\bar{G}(x)]\upsilon(dy). \end{aligned}$$

It remains to show that for every $f \in \mathcal{F}$, Af is bounded. Let r be the failure rate corresponding to $\bar{G}$, and for $x, y \in R_+$ we define

$$h(x, y) = \bar{G}^{-1}(x)[f(x+y)\bar{G}(x+y) - f(x)\bar{G}(x)].$$

Note that $\sup_{x,y} | h(x, y) | \leq 2\| f \|$, where $\| \ \|$ is the sup norm. Let $N = \| f \| \| r \| + \| f^{'} \|$, then

$$\begin{aligned} h(x, y) &= \bar{G}^{-1}(x) \int_{x}^{x+y} [f(u)\bar{G}^{'}(u) + f^{'}(u)\bar{G}(u)]du \\ &\leq y(\| f \| \| r \| + \| f^{'} \|) \\ &= yN. \end{aligned}$$

Therefore,

$$\begin{aligned} \| Af \| &= \| \int_{B} h(x, y)\upsilon(dy) \| \\ &\leq N \int_{0}^{1} y\upsilon(dy) + 2\| f \| \int_{1}^{\infty} \upsilon(dy) \\ &< \infty, \end{aligned}$$

where the last inequality follows since υ is the Lévy measure of the subordinator. ■

We note that the absolute continuity requirement for $\bar{G}$ on B can be dropped if the Lévy measure is finite.

Lemma 2.16 Let Ψ be the class of stopping times defined in Proposition 2.11, $\beta = \inf \varphi(\tau)$, where the infimum is taken over all $\tau \in \Psi$ for which $E[\tau] < \infty$, and assume that $\beta > 0$. For $x \geq 0$, define $c_2(x) = c - c_1(x)$. The following two problems are equivalent, in the sense that they have the same solution:

(P_1) Minimize $\varphi(\tau)$ over all $\tau \in \Psi$ for which $E[\tau] < \infty$.

(P_2) Maximize $\phi_\tau = \beta E[\tau] + E[c_2(Z_\tau)]$ over all $\tau \in \Psi$, for which $E[\tau] < \infty$.

Proof The proof is immediate upon realizing that, for every $\tau \in \Psi$,

$$\varphi(\tau) = \frac{c - E[c_2(Z_\tau), \tau < \zeta]}{E[\tau]} \tag{2.8}$$

$$= \frac{c - E[c_2(Z_\tau)]}{E[\tau]},$$

where the last equation follows since $c_2(x) = 0$ on $\bar{B}$. We omit the straightforward details. ■

The following theorem illustrates the fact that under suitable conditions on the cost functionals and $\bar{G}$, the optimal replacement policy that solves (P_1) of Lemma 2.16 is a control-limit policy.

Theorem 2.17 Assume that

(i) $E[\hat{\zeta}] < \infty$.

(ii) On B: $\bar{G}$ has a bounded failure rate, c_2 is differentiable, and c_2' is bounded.

(iii) the function $\beta + Ac_2(x)$ crosses the x-axis at most once, and if once, then the crossing is from above.

Let $a = inf\{z : \beta + Ac_2(z) \leq 0\}$, and $\hat{\tau}^* = \inf\{t \geq 0 : \hat{Z}_t \geq a\}$. Then, $\kappa(\hat{\tau}^*)$ is the solution to (P_1).

Proof From Proposition 2.11, it suffices to show that $\hat{\tau}^*$ is the solution to (P_2) for the stationary process $\hat{Z}$. From the definition of $\hat{\tau}^*$, we have $\hat{\tau}^* \in \hat{\Psi}$. From (i) it follows that $E[\hat{\tau}^*] < \infty$, and for any $\hat{\tau} \in \hat{\Psi}$, $E[\hat{\tau}] < \infty$. Using (ii), and (iii), from Theorem 2.14, for each $\hat{\tau} \in \hat{\Psi}$, we have

$$\phi_{\hat{\tau}^*} - \phi_{\hat{\tau}} = E[\int_0^{\hat{\tau}^*} Af(\hat{Z}_t)dt] - E[\int_0^{\hat{\tau}} Af(\hat{Z}_t)dt].$$

We express the right-hand side of the last equation as follows:

$$E[\int_{\hat{\tau}}^{\hat{\tau}^*} Af(\hat{Z}_t)dt\mathbf{I}_{\{\hat{\tau}\leq\hat{\tau}^*\}}] - E[\int_{\hat{\tau}^*}^{\hat{\tau}} Af(\hat{Z}_t)dt\mathbf{I}_{\{\hat{\tau}^*\leq\hat{\tau}\}}].$$

We note that the first term is positive by the definition of $\hat{\tau}^*$. From the definition of $\hat{\tau}^*$, (iv), and the fact that the process $\hat{Z}$ is increasing, we have $E[\int_{\hat{\tau}^*}^{\hat{\tau}} Af(\hat{Z}_t)dt\mathbf{I}_{\{\hat{\tau}^*\leq\hat{\tau}\}}] \leq 0$. Our assertion is thus proven. ■

The following theorem shows how the long-run average cost (using a control-limit policy) can be computed, with the help of the potential of the process $\hat{X}$, and the distribution of $\hat{X}_{\tau_x}$, where for any $x \in R_+$, $\tau_x = \inf\{t \geq 0 : \hat{X}_t \geq x\}$.

Theorem 2.18 Let $\hat{\mathbf{R}}$ be the potential of the process $\hat{X}$, and $\mathbf{R}(y) = \int_{[0,y)} \bar{G}(z)\hat{\mathbf{R}}(dz)$, $y \in R_+$. Let $c^* = c_2 \times \bar{G}$, and $\hat{\tau}_x = \inf\{t \geq 0 : \hat{Z}_t \geq x\}$. Denote the long-run average cost using $\hat{\tau}_x$ by $\psi(x)$, then

$$\psi(x) = \frac{c - E(c^*(\hat{X}\tau_x)}{\mathbf{R}(x)}.$$

Proof From (2.8), it suffices to show that $E[\hat{\tau}_x] = \mathbf{R}(x)$, and $E[c_2(Z_{\hat{\tau}_x}), \hat{\tau}_x < \zeta] = E[c^*(\hat{X}_{\tau_x})]$. We have

$$\begin{aligned}
E[\hat{\tau}_x] &= \int_0^\infty P\{\hat{\tau}_x > t\}dt \\
&= \int_0^\infty P\{\hat{Z}_t < x\}dt \\
&= \int_0^\infty P\{\hat{X}_t < x, t < \zeta\}dt \\
&= E\int_0^\infty \mathbf{I}_{\{\hat{X}_t<x\}}\mathbf{I}_{\{Y>\hat{X}_t\}} dt \\
&= E\int_0^\infty \mathbf{I}_{\{\hat{X}_t<x\}} \bar{G}(\hat{X}_t)dt
\end{aligned}$$

$$= \int_{[0,x)} \bar{G}(z)\hat{\mathbf{R}}(dz),$$

where the third equation follows from the definition of the process $\hat{Z}$, the fourth equation follows by first conditioning on $\hat{X}_t$ and since Y is assumed to be independent of $\hat{X}$, the last equation follows from (1.33).

Note that $\hat{\tau}_x = \tau_x \wedge \zeta$, hence

$$\begin{aligned} E[c_2(\hat{Z}_{\hat{\tau}_x}), \hat{\tau}_x < \zeta] &= E[c_2(\hat{X}_{\tau_x}), \tau_x < \zeta] \\ &= E[c_2(\hat{X}_{\tau_x})\bar{G}(\hat{X}_{\tau_x})] \\ &= E[c^*(\hat{X}_{\tau_x})], \end{aligned}$$

where the third equation follows from the second equation upon conditioning on $\hat{X}_{\tau_x}$ and since the threshold Y is assumed to be independent of $\hat{X}$. Our assertion is thus proven. ■

In [22], a similar maintenance model is considered assuming imperfect maintenance actions, that is, each maintenance does not restore the system to good as new condition. The authors give conditions that insure that the device has decreasing short-run availability.

2.5 One-Level Control-Limit Maintenance Policies for Non-Continuously Monitored Degradable Systems

In this case, we assume that the deterioration is only observed at inspection times $T_n, n = 1, \ldots$. When a failure is detected upon inspection, a *corrective maintenance* operation replaces the failed system with a new and identical one. If upon inspection, the monitored deterioration exceeds level M and is less than threshold Y, a *preventive maintenance* operation replaces the system with a new and identical one. When upon inspection, the monitored condition is less than M, nothing is done and a new inspection date is chosen. The choice of inspection times and M will influence the performance of the maintenance policy. The maintenance policy is driven by the knowledge of system state at times of inspection. If at any inspection time the deterioration is less than M, the system is left alone. Thus, at any inspection time, the system is either left alone, maintained correctively or preventively.

Specifically, we assume that a system is subject to degradation over time. The degradation process is assumed to be a nonhomogeneous subordinator with zero drift, denoted by $Z = \{Z_t, t \geq 0\}$. The system has a nominal life distribution Y, with right tail probability $\bar{G}$ and once the degradation exceeds Y, the system fails. The states of the system are only observed by inspection, also failures are only detected by inspection. Let $\hat{T} = \{\hat{T}_n, n = 1, 2, ...\}$ be the times of successive inspections, $\hat{T}_0 = 0$; for $n \geq 0$ we denote $Z_{\hat{T}_n}$ by $\hat{X}_n$, and given $\hat{X}_n$ and $\hat{T}_n$ the time of the next inspection $\hat{T}_{n+1}$ is chosen using the rule: $\hat{T}_{n+1} = \hat{T}_n + W(\hat{X}_n)$, where for $x \in R_+$, $W(x)$ is a positive random variable. The system is replaced when a failure is detected (*Corrective Maintenance* (CM)), or once the observed degradation exceeds level M (*Preventive maintenance* (MP)). The system is as good as new after each maintenance (Corrective or Preventive). We assume that the maintenance operation has a negligible duration. Given $\hat{T}_n$ and $\hat{X}_n$, the following maintenance decision frame is adopted:

(1) If $\hat{X}_n < M \wedge Y$, the system is left unchanged, no maintenance is performed, and the next inspection takes place at time $\hat{T}_{n+1}$.
(2) If $M \leq \hat{X}_n < Y$, (the system is functioning but is subject to a high deterioration level) a preventive maintenance is performed at a cost c_1.
(3) If $\hat{X}_n \geq Y$ (the system failed) a corrective maintenance is performed at a cost $c_1 + k, k > 0$.
(4) If $M \leq \hat{X}_n < Y$ or $\hat{X}_n \geq Y$, then the next inspection time is $\hat{T}_n + W(0)$.
(5) If at inspection the degradation level is y, a penalty cost is accrued at a rate $c(y)$ until the next inspection occurs.

We note that (4) above is a restatement of the assumption that each replacement is made with a new system with zero degradation (complete repair).

For $x \in R_+$, we let $m(x) = E[W(x)]$, where $m : R_+ \to R_+$ is assumed to be bounded, and we let F_x be the distribution function of $W(x)$.

Following the same argument used in Proposition 2.11, it suffices to deal with the case where the deterioration process is stationary, and we abuse the notation and denote it by Z. Since the process Z is a strong Markov process, with stationary independent increments, the process $(\hat{X}, \hat{T}) = (\hat{X}_n, \hat{T}_n, n = 1, 2, ...)$ is a Markov renewal process with state space $[0, \infty)$. For any $t, x \in R_+$, $A \in B(R_+)$, the associated semi-Markov kernel is given by

$$\begin{aligned} \hat{Q}(x, A, t) &= P\{\hat{X}_{n+1} \in A, \hat{T}_{n+1} - \hat{T}_n \leq t \mid \hat{X}_n = x, \hat{T}_n\} \\ &= P\{Z_{W(x)} \in A - x, W(x) \leq t\} \\ &= \int_{[0,t]} P\{Z_s \in A - x\} F_x(ds) \end{aligned} \tag{2.9}$$

For $n \geq 0$, we let

$$\hat{Q}^n(x, A, t) = P_x\{\hat{X}_n \in A, \hat{T}_n \leq t\}.$$

It follows that, for $n \geq 1$,

$$\hat{Q}^n(x, A, t) = \int\int_{A\times[0.t]} \hat{Q}^{n-1}(y, A, t-s)\hat{Q}(x, dy, ds),$$

where

$$\hat{Q}^0(x, A, t) = \begin{cases} 1, & \text{if } x \in A, \\ 0, & \text{if } x \notin A. \end{cases}$$

The Markov renewal kernel of the process $(\hat{X}, \hat{T})$ is defined to be

$$\hat{R}(x, A, t) = \sum_{n=0}^{\infty} \hat{Q}^n(x, A, t).$$

For $\alpha, x \in R_+$, and $A \in B(R_+)$, we define

$$\hat{R}_\alpha(x, A) = \int_0^\infty e^{-\alpha t}\hat{R}(x, A, dt).$$

It follows that

$$\hat{R}_\alpha(x, A) = \sum_{n=0}^{\infty} \hat{Q}^n_\alpha(x, A),$$

where

$$\hat{Q}^n_\alpha(x, A) = \int_0^\infty e^{-\alpha t}\hat{Q}^n(x, A, dt).$$

We define the semi-Markov process, $\hat{Y}$, induced by $(\hat{X}, \hat{T})$ as follows: For $t \geq 0$,

$$\hat{Y}_t = \{\hat{X}_n, \hat{T}_n \leq t < \hat{T}_{n+1}, \ n = 0, 1, ...\}.$$

Let ς be the life time of the device, that is

$$\varsigma = \inf\{t \geq 0 : \hat{Y}_t \geq Y\},$$

and

$$\hat{T}_M = \inf\{t \geq 0 : \hat{Y}_t \geq M\}.$$

Let T_M be the random variable denoting the maintenance time (the minimum of the corrective maintenance and preventative maintenance times), that is

$$T_M = \hat{T}_M \wedge \varsigma\,.$$

For $z, \alpha \in R_+$, we define $h_\alpha(z) = \frac{1}{\alpha} E(1 - e^{-\alpha W(z)})$. For $\alpha \in R_+$, we let $\hat{U}^{\alpha}$ be the α-potential of the process $\hat{Y}$. Observe that, for $x \in R_+$, and $A \subset R_+$

$$\begin{aligned}
\hat{U}_\alpha(x, A) &= E_x[\int_0^\infty e^{-\alpha t} \mathbf{I}_{\{\hat{Y}_t \in A\}} dt] \\
&= E_x[\int_0^\infty e^{-\alpha t} \sum_{n=0}^{\infty} (\mathbf{I}_{\{\hat{X}_n \in A\}} \mathbf{I}_{\{\hat{T}_n \leq t < \hat{T}_{n+1}\}}) dt] \\
&= \frac{1}{\alpha} \sum_{n=0}^{\infty} E_x[\mathbf{I}_{\{\hat{X}_n \in A\}} (e^{-\alpha \hat{T}_n} - e^{-\alpha \hat{T}_{n+1}})] \\
&= \frac{1}{\alpha} \sum_{n=0}^{\infty} E_x[E_x[\mathbf{I}_{\{\hat{X}_n \in A\}} (e^{-\alpha \hat{T}_n} - e^{-\alpha \hat{T}_{n+1}}) \mid \hat{X}_n, \hat{T}_n]] \\
&= \frac{1}{\alpha} \sum_{n=0}^{\infty} E_x[\mathbf{I}_{\{\hat{X}_n \in A\}} e^{-\alpha \hat{T}_n} E_{\hat{X}_n} [1 - e^{-\alpha \hat{T}_1}]] \\
&= \frac{1}{\alpha} \sum_{n=0}^{\infty} E_x[\mathbf{I}_{\{\hat{X}_n \in A\}} e^{-\alpha \hat{T}_n} E(1 - e^{-\alpha W(\hat{X}_n)})] \\
&= \sum_{n=0}^{\infty} E_x[\mathbf{I}_{\{\hat{X}_n \in A\}} e^{-\alpha \hat{T}_n} h_\alpha(\hat{X}_n)] \\
&= \sum_{n=0}^{\infty} \int_A \hat{Q}{}^n_\alpha(x, dy) h_\alpha(y) \\
&= \int_A \hat{R}_\alpha(x, dy) h_\alpha(y), \qquad (2.10)
\end{aligned}$$

where the third and last equations follow from *Fubini's Theorem.*

Since after each maintenance operation the system is as good as new, then after such operation the deterioration process is brought back to level zero. Let L denote the n such that $\hat{T}_n$ is the failure time of the system, that is $L = \inf\{n : \hat{T}_n = \varsigma\}$, then the failure time $\varsigma = \hat{T}_L$. For $t \geq 0$, $n \geq 0$, let

$$Y_t = \{\hat{Y}_t, t < \varsigma\},$$

$$X_n = \{\hat{X}_n, n < L\},$$

and

$$T_n = \{\hat{T}_n, n \leq L\}.$$

We note that the process Y is a semi-Markov process, obtained by killing the process $\hat{Y}$ at the time of first failure, and $(X, T) = \{(X_n, T_n), n \geq 0\}$ is its Markov renewal process. For $x \in R_+$, $A \in \sigma([x, \infty))$, let $Q^n(x, A, t) = P_x\{X_n \in A, T_n \leq t \mid X_0 = x\}$. Then, $Q(x, A, t) \stackrel{def}{=} Q^1(x, A, t)$ is the semi-Markov kernel associated with (X, T).

Lemma 2.19 For $x, t \in R_+$, $A \in \beta([x, \infty))$,

$$Q(x, A, t) = \int_A \hat{Q}(x, dy, t)\frac{\bar{G}(y)}{\bar{G}(x)}$$

Proof For $n \geq 1$, let F_n be the sigma algebra generated by $\{(\hat{X}_k, \hat{T}_k), 0 \leq k \leq n\}$. Then,

$$\begin{aligned}
Q(x, A, t) &= P\{X_1 \in A, T_1 \leq t \mid X_0 = x\} \\
&= P\{\hat{X}_1 \in A, \hat{T}_1 \leq t, L > 1 \mid \hat{X}_0 = x, L > 0\} \\
&= P\{\hat{X}_1 \in A, \hat{T}_1 \leq t, \hat{X}_1 < Y \mid \hat{X}_0 = x, \hat{X}_0 < Y\} \\
&= [\bar{G}(x)]^{-1} P\{\hat{X}_1 \in A, \hat{T}_1 \leq t, \hat{X}_1 < Y \mid \hat{X}_0 = x\} \\
&= [\bar{G}(x)]^{-1} E_x[\mathbf{I}_{\{\hat{X}_1 < Y\}} \mathbf{I}_{\{\hat{X}_1 \in A\}} \mathbf{I}_{\{\hat{T}_1 \leq t\}}] \\
&= [\bar{G}(x)]^{-1} E_x[E_x[\mathbf{I}_{\{\hat{X}_1 < Y\}} \mathbf{I}_{\{\hat{X}_1 \in A\}} \mathbf{I}_{\{\hat{T}_1 \leq t\}} \mid F_1]] \\
&= [\bar{G}(x)]^{-1} E_x[\mathbf{I}_{\{\hat{X}_1 \in A\}} \mathbf{I}_{\{\hat{T}_1 \leq t\}} E_x[\mathbf{I}_{\{\hat{X}_1 < Y\}} \mid F_1]] \\
&= E_x[\frac{\bar{G}(\hat{X}_1)}{\bar{G}(x)} \mathbf{I}_{\{\hat{X}_1 \in A\}} \mathbf{I}_{\{\hat{T}_1 \leq t\}}]
\end{aligned}$$

$$= \int_A \hat{Q}(x, dy, t) \frac{\bar{G}(y)}{\bar{G}(x)},$$

where the eighth equality follows because the threshold variable Y is independent of $(\hat{X}, \hat{T})$, and our assertion is proven. ■

Proposition 2.20 Let R, U^{α}, be the Markov renewal kernel associated with the process (X, T), and the α-potential of the process Y, respectively. Then,

(i) For $x, t \in R_+$, $A \subset R_+$,

$$R(x, A, t) = \int_A \hat{R}(x, dy, t) \frac{\bar{G}(y)}{\bar{G}(x)}.$$

(ii) For $x \in R_+$, $A \subset R_+$,

$$U_{\alpha}(x, A) = \int_A R_{\alpha}(x, dy) h_{\alpha}(y).$$

Proof (i) For $n \geq 1$, we let

$$Q^n(x, dy, t) = P\{X_n \in A, T_n \leq t \mid X_0 = x\}.$$

Following an argument similar to the one used in proving Lemma 2.19 it follows that, for $n \geq 0$,

$$Q^n(x, A, t) = \int_A \hat{Q}^n(x, dy, t) \frac{\bar{G}(y)}{\bar{G}(x)}.$$

Hence,

$$\begin{aligned} R(x, A, t) &= \sum_{n=0}^{\infty} Q^n(x, A, t) \\ &= \sum_{n=0}^{\infty} \int_A \hat{Q}^n(x, dy, t) \frac{\bar{G}(y)}{\bar{G}(x)} \\ &= \int_A \hat{R}(x, dy, t) \frac{\bar{G}(y)}{\bar{G}(x)}, \end{aligned}$$

where the last equation follows from the monotone convergence theorem. This proves the first assertion of the proposition.

(ii) Note that

$$P_x\{Y_t \in A\} = P_x\{\hat{Y}_t \in A, t < \varsigma\}$$

$$
\begin{aligned}
&= P_x\{\hat{Y}_t \in A, \hat{Y}_t < Y\} \\
&= E_x[\mathbf{I}_{\{\hat{Y}_t \in A\}} \mathbf{I}_{\{\hat{Y}_t < Y\}}] \\
&= E_x[\mathbf{I}_{\{\hat{Y}_t \in A\}} E_x[\mathbf{I}_{\{\hat{Y}_t < Y\}} \mid \hat{Y}_t]] \\
&= \frac{1}{\bar{G}(x)} E_x[\bar{G}(\hat{Y}_t)\mathbf{I}_{\{\hat{Y}_t \in A\}}],
\end{aligned}
$$

where the last equation follows since the threshold random variable Y is independent of the process $\hat{Y}_t$.

Thus,

$$
\begin{aligned}
U_\alpha(x, A) &= E_x[\int_0^\infty e^{-\alpha t} \mathbf{I}_{\{Y_t \in A\}} dt] \\
&= \int_0^\infty e^{-\alpha t} P_x\{Y_t \in A\} \\
&= \int_A \hat{U}_\alpha(x, dy) \frac{\bar{G}(y)}{\bar{G}(x)} \\
&= \int_A \hat{R}_\alpha(x, dy) h_\alpha(y) \frac{\bar{G}(y)}{\bar{G}(x)} \\
&= \int_A R_\alpha(x, dy) h_\alpha(y),
\end{aligned}
$$

where the fourth equation follows from (2.10) and the last equation follows from (i) of this proposition. ■

We note that for $x \in R_+$, $R(x, [x, \infty), \infty)$ is the expected number of inspections before the first failure occurs, given $Z_0 = x$.

The following theorem gives Laplace transform of the distribution function of the time of first replacement.

Theorem 2.21 For $x < M$,

$$
E_x[e^{-\alpha T_M}] = 1 - \alpha \int_{[x, M]} R_\alpha(x, dy) h_\alpha(y)
$$

Proof Write

$$U_\alpha \mathbf{I}_{[x,M]}(x) = E_x[\int_0^\infty e^{-\alpha t}\mathbf{I}_{[x,M]}(Y_t)dt]$$
$$= E_x[\int_0^{T_M} e^{-\alpha t}\mathbf{I}_{[x,M]}(Y_t)dt] + E_x[\int_{T_M}^\infty e^{-\alpha t}\mathbf{I}_{[x,M]}(Y_t)dt]$$
$$= E_x[\int_0^{T_M} e^{-\alpha t}\mathbf{I}_{[x,M]}(Y_t)dt]$$

$$= E_x[\int_0^{T_M} e^{-\alpha t}dt]$$
$$= \frac{1}{\alpha}(1 - E_x[e^{-\alpha T_M}]).$$

Thus,

$$E_x[e^{-\alpha T_M}] = 1 - \alpha U_\alpha \mathbf{I}_{[x,M]}(x),$$

and our assertion follows from (ii) of Proposition 2.20. ■

Corollary 2.22 (i) For $x \in R_+$, and given $Z_0 = x$, the expected time till the first replacement is as follows:

$$E_x[T_M] = \int_{[x,M]} R(x, dy, \infty)m(y).$$

(ii) For $x \in R_+$, and given $Z_0 = x$, the following formula gives the mean of the failure time

$$E_x[\zeta] = \int_{[x,\infty)} R(x, dy, \infty)m(y)$$

Proof (i) As shown in the proof of Theorem 2.21, for $\alpha, x \in R_+$,

$$\frac{1}{\alpha}(1 - E_x[e^{-\alpha T_M}]) = U_\alpha \mathbf{I}_{[x,M]}(x).$$

Thus,

$$
\begin{aligned}
E_x[T_M] &= \lim_{\alpha\to 0} \frac{1}{\alpha}(1-E_x[e^{-\alpha t T_M}]) \\
&= \lim_{\alpha\to 0} \int_{[x,M]} R_\alpha(x,dy)h_\alpha(y) \\
&= \int_{[x,M]} \lim_{\alpha\to 0} [R_\alpha(x,dy)h_\alpha(y)] \\
&= \int_{[x,M]} R(x,dy,\infty)m(y)
\end{aligned}
$$

where the third equation follows from the *Lebesgue Monotone Convergence Theorem,* since

$$
\begin{aligned}
\lim_{\alpha\to 0} h_\alpha(y) &= \lim_{\alpha\to 0} \int_{R_+} e^{-\alpha t} P_y(T_1 > t) \\
&= \int_{R_+} \lim_{\alpha\to 0} e^{-\alpha t} P_y(T_1 > t) \\
&= \int_{R_+} P_y(T_1 > t) \\
&= E_y(T_1) \\
&= m(y).
\end{aligned}
$$

This proves our first assertion.

(ii) The proof of (ii) follows from (i) since $\lim_{M\to\infty} T_M = \zeta$ almost everywhere. ■

The following gives sufficient conditions for $E_x[\zeta] < \infty$.

Proposition 2.23 Assume that for each $x \in R_+$, $W(x) = W$, where W is a positive random variable, $\bar{G}$ is NBU, and $E[\bar{G}(Z_W)] \neq 1$. Then, for each $x \in R_+$, $E_x[\zeta] < \infty$.

Proof We note that the first assumption implies that the times between two successive inspections are independent identically distributed random variables. Thus, for $n \geq 1$,

$$\hat{X}_n = \sum_{i=1}^{n} Y_i,$$

where $\{Y_i, i \geq 1\}$ is a sequence of independent identically random variables such that, for each Borel set $A \subset R_+$

$$P\{Y_i \in A\} = P\{Z_W \in A\}.$$

Since $\bar{G}$ is NBU, then for any $n \geq 1$, and $x_1, .., x_n \in R_+$,

$$\bar{G}(x_1 + \cdots + x_n) \leq \prod_{i=1}^{n} \bar{G}(x_i).$$

Let $m = E(W)$. From (ii) of Corollary 2.22, we have

$$\begin{aligned} E_x[\zeta] &= m \int_{[x,\infty)} R(x, dy, \infty) \\ &= mR(x, R_+, \infty). \end{aligned}$$

For each $x \in R_+$, and $n \geq 1$,

$$\begin{aligned} Q^n(x, R_+, \infty) &= \int_{[x,\infty)} \hat{Q}^n(x, dy, \infty)\frac{\bar{G}(y)}{\bar{G}(x)} \\ &= \int_{[x,\infty)} P_x\{\hat{X}_n \in dy)\frac{\bar{G}(y)}{\bar{G}(x)} \\ &= \int_{[x,\infty)} P\{\hat{X}_n \in dy - x)\frac{\bar{G}(y)}{\bar{G}(x)} \\ &= \int_{[0,\infty)} P\{\hat{X}_n \in dy)\frac{\bar{G}(x+y)}{\bar{G}(x)} \\ &\leq \int_{[0,\infty)} P\{\hat{X}_n \in dy)\bar{G}(y) \\ &= E[\bar{G}(\hat{X}_n)] \\ &= E[\bar{G}(\sum_{i=1}^{n} Y_i)] \end{aligned}$$

$$\leq \prod_{i=1}^{n} E[\bar{G}(Y_i)]$$
$$= (E[\bar{G}(Y_1)])^n.$$

Thus,

$$\begin{aligned} R(x, R_+, \infty) &= \sum_{i=0}^{n} Q^n(x, R_+, \infty) \\ &\leq \sum_{i=0}^{n} (E[\bar{G}(Y_1)])^n \\ &= \frac{1}{1 - E[\bar{G}(Y_1)]} \\ &< \infty. \end{aligned}$$

This finishes the proof. ■

The following gives a sufficient condition for $E_x[T_M] < \infty$.

Proposition 2.24 Assume that for each $x \in R_+$, $W(x) = W$, where W is a positive random variable. Then, for each $x \in R_+$, $E_x[T_M] < \infty$.

Proof From (i) of Corollary 2.22, we have

$$\begin{aligned} E_x[T_M] &= m \int_{[x,M]} R(x, dy, \infty) \\ &= m R(x, [x, M], \infty), \end{aligned}$$

where $m = E(W)$.

Let $\{Y_i, i \geq 1\}$ and $\{\hat{X}_n, n \geq 1\}$ be as defined in Proposition 2.23; following an argument similar to the one used in proving this proposition, for $x \leq M$ and $n \geq 1$, we have

$$\begin{aligned} Q^n(x, [x, M], \infty) &= \int_{[x,M]} \hat{Q}^n(x, dy, \infty) \frac{\bar{G}(y)}{\bar{G}(x)} \\ &= \int_{[x,M]} P_x\{\hat{X}_n \in dy\} \frac{\bar{G}(y)}{\bar{G}(x)} \\ &\leq P\{\hat{X}_n \leq M - x\}. \end{aligned}$$

Let F denote the distribution function of the random variable Y_i. Then, for $n \geq 1$,

$$\begin{aligned} Q^n(x, [x, M], \infty) &= P\{\hat{X}_n \leq M - x\} \\ &\leq P\{\hat{X}_n \leq M\} \\ &= F^{(n)}(M), \end{aligned}$$

where for $n \geq 1$, $F^{(n)}$ is the nth convolution of F.

We claim that for $n \geq 1$, $F^{(n)}(M) \leq (F(M))^n$. This is proven by induction on n, since

$$\begin{aligned} F^{(2)}(M) &= \int_{[0,M]} F(M-y)F(dy) \\ &\leq F(M) \int_{[0,M]} F(dy) \\ &= (F(M))^2. \end{aligned}$$

Now assuming that our claim is true for a given n, then

$$\begin{aligned} F^{(n+1)}(M) &= \int_{[0,M]} F^{(n)}(M-y)F(dy) \\ &\leq F^{(n)}(M) \int_{[0,M]} F(dy) \\ &\leq (F(M))^n \int_{[0,M]} F(dy) \\ &= (F(M))^{n+1}, \end{aligned}$$

this finishes the proof of the claim.

From the properties of the process Z it follows that $F(M) < 1$. Hence

$$\begin{aligned} R(x, [x, M], \infty) &= \sum_{n=0}^{\infty} Q^n(x, [x, M], \infty) \\ &\leq \sum_{n=0}^{\infty} F^{(n)}(M) \\ &\leq \sum_{n=0}^{\infty} (F(M))^n \\ &= \frac{1}{1 - F(M)} \\ &< \infty, \end{aligned}$$

and our assertion is proven. ■

Proposition 2.25 For any $\alpha, x \in R_+$,

$$E_x[e^{-\alpha T_M}, T_M = \zeta] = \int_{[x,M]} R_\alpha(x, dy)[\hat{Q}_\alpha(y, R_+) - Q_\alpha(y, R_+)]$$

Proof For $n \geq 1$, we let G_n be the sigma algebra generated by $\{(X_k, T_k), 1 \leq k \leq n\}$. Then

$$\begin{aligned}
E_x[e^{-\alpha T_M}, T_M = \zeta] = E_x[e^{-\alpha\zeta}, Y_{\zeta-} < M] & \\
&= E_x[e^{-\alpha\zeta}, Y_{\zeta-} < M, X_1 = \infty] \\
&\quad + E_x[e^{-\alpha\zeta}, Y_{\zeta-} < M, X_1 < \infty] \\
&= E_x[e^{-\alpha T_1}, X_0 < M, X_1 = \infty] \\
&\quad + E_x[e^{-\alpha\zeta}, Y_{\zeta-} < M, X_1 < \infty] \\
&= \mathbf{I}_{[0,M)}(x) E_x[e^{-\alpha T_1}, X_1 = \infty] \\
&\quad + E_x[E_x[e^{-\alpha\zeta}, Y_{\zeta-} < M, X_1 < \infty \,|\, G_1]] \\
&= \mathbf{I}_{[0,M)}(x)\{E_x[e^{-\alpha T_1}] - E_x[e^{-\alpha T_1}, X_1 < \infty]\} \\
&\quad + \int_0^\infty Q_\alpha(x, dy) E_y[e^{-\alpha T_M}, T_M = \zeta] \\
&= \mathbf{I}_{[0,M)}(x)[\hat{Q}_\alpha(x, R_+) - Q_\alpha(x, R_+)] \\
&\quad + \int_0^\infty Q_\alpha(x, dy) E_y[e^{-\alpha T_M}, T_M = \zeta].
\end{aligned}$$

Using Theorem 10.3.6 and Proposition 10.3.14 of [23], it follows that the unique solution of the last equation is

$$E_x[e^{-\alpha T_M}, T_M = \zeta] = \int_{[x,M]} R_\alpha(x, dy)[\hat{Q}_\alpha(y, R_+) - Q_\alpha(y, R_+)],$$

and our assertion is proven. ■

Corollary 2.26 For any $x \in R_+$,

$$P_x\{T_M = \zeta\} = \int_{[x,M]} R(x, dy, \infty)[1 - Q(y, R_+, \infty)].$$

Proof Write

$$
\begin{aligned}
P_x\{T_M = \zeta\} &= E_x[\lim_{\alpha\to 0} e^{-\alpha T_M}, T_M = \zeta] \\
&= \lim_{\alpha\to 0} E_x[e^{-\alpha T_M}, T_M = \zeta] \\
&= \lim_{\alpha\to 0} \int_{[x,M]} R_\alpha(x, dy)[\hat{Q}_\alpha(y, R_+) - Q_\alpha(y, R_+)] \\
&= \int_{[x,M]} \lim_{\alpha\to 0} R_\alpha(x, dy) \lim_{\alpha\to 0} [\hat{Q}_\alpha(y, R_+) - Q_\alpha(y, R_+)]\} \\
&= \int_{[x,M]} R(x, dy, \infty)[1 - Q(y, R_+, \infty)],
\end{aligned}
$$

where the second, fourth, and last equations follow from the *Lebesgue Dominated Convergence Theorem,* while the third equation follows from Proposition 2.25. ■

Lemma 2.27 Let g be a bounded measurable function on R_+. Then for $x \in R_+$,

$$
E_x[\int_0^{T_M} e^{-\alpha t} g(Y_t)dt] = \int_{[x,M]} R_\alpha(x, dy)h_\alpha(y)g(y).
$$

Proof Write

$$
\begin{aligned}
E_x[\int_0^{T_M} e^{-\alpha t} g(Y_t)dt] &= E_x[\int_0^{\infty} e^{-\alpha t} \mathbf{I}_{\{T_M > t\}} g(Y_t)dt] \\
&= E_x[\int_0^{\infty} e^{-\alpha t} \mathbf{I}_{\{Y_t < M\}} g(Y_t)dt] \\
&= \int_{[x,M]} U_\alpha(x, dy)g(y) \\
&= \int_{[x,M]} R_\alpha(x, dy)h_\alpha(y)g(y),
\end{aligned}
$$

where the last equation follows from (ii) of Proposition 2.20. ■

Corollary 2.28 For $x \le M$

$$E_x\Big[\int_0^{T_M} g(Y_t)dt\Big] = \int_{[x,M]} R(x, dy, \infty)m(y)g(y).$$

Proof Write

$$\begin{aligned}
E_x\Big[\int_0^{T_M} g(Y_t)dt\Big] &= E_x\Big[\int_0^{T_M} \lim_{\alpha\to 0} e^{-\alpha t} g(Y_t)dt\Big] \\
&= \lim_{\alpha\to 0} E_x\Big[\int_0^{T_M} e^{-\alpha t} g(Y_t)dt\Big] \\
&= \lim_{\alpha\to 0} \int_{[x,M]} R_\alpha(x, dy)h_\alpha(y)g(y) \\
&= \int_{[x,M]} \lim_{\alpha\to 0} [R_\alpha(x, dy)h_\alpha(y)]g(y) \\
&= \int_{[x,M]} R(x, dy, \infty)m(y)g(y),
\end{aligned}$$

where the second and fourth equations follow from the *Monotone Convergence Theorem*, and the third equation follows from Lemma 2.27. ■

Denote Y by $\overset{(1)}{Y}$, let $\{\overset{(n)}{Y},\ n \ge 2\}$ be a sequence of independent copies of the process $\overset{(1)}{Y}$, with the exception that each starts at zero (each is independent of $\overset{(1)}{Y}$), and $\overset{(n)}{T}_M$ be the replacement time corresponding to $\overset{(n)}{Y}$. DenoteT_M by $\overset{(1)}{T}_M$. For $n \ge 1$, let $S_n = \sum_{k=1}^{n} \overset{(k)}{T}_M$. We define the generic process $Z = \{Z_t, t \ge 0\}$ as follows:

$$Z_t = \sum_{n=1}^{\infty} \overset{(n)}{Y}_{t-S_{n-1}} \mathbf{I}_{\{S_{n-1} \le t << S_n\}}, \quad Z_{S_n} = 0 \quad \text{for} \quad n \ge 1.$$

It is clear that the process Z describes the deterioration over the infinite horizon, when replacements are made at the successive replacement times. Furthermore, it is a delayed regenerative process with regeneration times $\{S_n, n \ge 1\}$.

Theorem 2.29 Let $g : R_+ \to R_+$, be a bounded measurable function. For $x \in R_+$, let $c_\alpha(x, M) = E_x[\int_0^{T_M} e^{-\alpha t} g(Y_t)dt]$, and $C_\alpha(x) = E_x[\int_0^{\infty} e^{-\alpha t} g(Z_t)dt]$. Then,

$$C_\alpha(x) = c_\alpha(x, M) + \frac{E_x[e^{-\alpha T_M}]c_\alpha(0, M)}{1 - E_0[e^{-\alpha T_M}]}. \tag{2.11}$$

Proof For $t \geq 0$, let F_t be the sigma algebra generated by $\{Z_s, s \leq t\}$, then

$$\begin{aligned}
C_\alpha(x) &= E_x[\sum_{n=1}^{\infty} \int_{S_{n-1}}^{S_n} e^{-\alpha t} g(\overset{(n)}{Y}_{t-S_{n-1}})dt] \\
&= E_x[\int_0^{T_M} e^{-\alpha t} g(Y_t)dt] + E_x[\sum_{n=2}^{\infty} \int_{S_{n-1}}^{S_n} e^{-\alpha t} g(\overset{(n)}{Y}_{t-S_{n-1}})dt] \\
&= c_\alpha(x, M) + \sum_{n=2}^{\infty} E_x[\int_{S_{n-1}}^{S_n} e^{-\alpha t} g(\overset{(n)}{Y}_{t-S_{n-1}})dt] \\
&= c_\alpha(x, M) + \sum_{n=2}^{\infty} E_x[e^{-\alpha S_{n-1}} \int_0^{\overset{(n)}{T}_M} e^{-\alpha t} g(\overset{(n)}{Y}_t)dt] \\
&= c_\alpha(x, M) + \sum_{n=1}^{\infty} E_x[E_x[e^{-\alpha S_n} \int_0^{\overset{(n+1)}{T}_M} e^{-\alpha t} g(\overset{(n+1)}{Y_t})dt \mid F_{S_n}]] \\
&= c_\alpha(x, M) + \sum_{n=1}^{\infty} E_x[e^{-\alpha S_n} E_0[\int_0^{T_M} e^{-\alpha t} g(Y_t)dt]] \\
&= c_\alpha(x, M) + c_\alpha(0, M) \sum_{n=1}^{\infty} E_x[e^{-\alpha S_n}] \\
&= c_\alpha(x, M) + c_\alpha(0, M) \sum_{n=1}^{\infty} E_x[E_x[e^{-\alpha S_n} \mid F_{S_1}]] \\
&= c_\alpha(x, M) + c_\alpha(0, M) \sum_{n=1}^{\infty} E_x[e^{-\alpha S_1} E_0[e^{-\alpha S_{n-1}}]] \\
&= c_\alpha(x, M) + c_\alpha(0, M) E_x[e^{-\alpha T_M}] \sum_{n=0}^{\infty} E_0[e^{-\alpha S_n}]
\end{aligned}$$

$$= c_\alpha(x, M) + \frac{E_x[e^{-\alpha T_M}]c_\alpha(0, M))}{1 - E_0[e^{-\alpha T_M}]}$$

where the third equation follows from *Fubini's Theorem*. ■

We now turn our attention to computing the total discounted cost. According to the maintenance decision form adopted, c_1 is the cost of a preventive maintenance, $c_1 + k$ is the cost of a corrective maintenance, and $c(.)$ is the rate of the penalty cost. A cycle is defined as the time between two successive replacement times. We note that the discounted cost during any cycle is equal to

$$\{[c_1 e^{-\alpha T_M} + \int_0^{T_M} c(Y_t)dt],\ T_M < \zeta\} + \{[(c_1 + k)e^{-\alpha T_M} + \int_0^{T_M} e^{-\alpha t} c(Y_t)dt],\ T_M = \zeta\}.$$

Thus, the expected discounted cost during the first cycle, when the deterioration at time zero being equal to x is given as follows:

$$c_\alpha(x, M) = c_1 E_x[e^{-\alpha T_M}] + kE_x[e^{-\alpha T_M},\ T_M = \zeta] + E_x[\int_0^{T_M} e^{-\alpha t} c(Y_t)dt]. \tag{2.12}$$

Using Theorem 2.21, Proposition 2.25, and Lemma 2.27, it follows that

$$c_\alpha(x, M) = c_1 + \int_{[x,M]} R_\alpha(x, dy)([h_\alpha(y)c(y) - \alpha c_1] + k[\hat{Q}_\alpha(y, R_+) - Q_\alpha(y, R_+)]). \tag{2.13}$$

Assuming that $Z_0 = 0$ and using (2.11), the total discounted cost is given by

$$C_\alpha(0) = \frac{c_1 + \int_{[0,M]} R_\alpha(0, dy)([h_\alpha(y)c(y) - \alpha c_1] + k[\hat{Q}_\alpha(y, R_+) - Q_\alpha(y, R_+)])}{\alpha \int_{[x,M]} R_\alpha(0, dy)h_\alpha(y)}. \tag{2.14}$$

The following illustrates how the optimal value M^* that minimizes the total discounted cost is computed.

Theorem 2.30 Assume that $A \to R_\alpha(0, A)$ is absolutely continuous with respect to the Lebesgue measure on R_+, and

(i) $y \to h_\alpha(y)$ is decreasing,

(ii) $y \to [\hat{Q}_\alpha(y, R_+) - Q_\alpha(y, R_+)]$ is increasing,

(iii) $y \to c(y)$ is increasing and bounded.

Then the optimal value M^* is the unique solution of the integral equation

$$c_1 = \int_{[0,M]} R_\alpha(0, dy)\{h_\alpha(y)[c(M) - c(y)] + k(\frac{h_\alpha(y)}{h_\alpha(M)}[\hat{Q}_\alpha(M, R_+) - Q_\alpha(M, R_+)] - [\hat{Q}_\alpha(y, R_+) - Q_\alpha(y, R_+)])\}. \quad (2.15)$$

If no solution exists, then the optimal policy is to wait until failure for replacement.

Proof Differentiating the right-hand side of (2.14) with respect to M, setting it equal to zero, and simplifying we obtain (2.15). By assumption, the integrand is positive, and hence (2.15) has at most one solution. ■

Remark 2.31 (i) If the random variable $W(x)$ has the same distribution regardless of the value of x, then $h_\alpha(y)$ is independent of y, and condition (i) of the last theorem is satisfied.

(ii) Under the same assumption in (i), and if $\bar{G}$ is IFR, then condition (ii) of the above theorem is satisfied.

To determine the optimal replacement policy using the long-run average cost criterion we proceed as follows: Denote the non discounted cost during a cycle, when the deterioration at the beginning of the cycle is zero, by $c(0, M)$. Letting $\alpha \to 0$, in (2.12), we have

$$c(0, M) = c_1 + kP\{T_M = \zeta\} + E[\int_0^{T_M} c(Y_t)dt].$$

Note that if $E[T_M] < \infty$, then $E[\int_0^{T_M} c(Y_t)dt] \leq \|c\| E[T_M] < \infty$. Using Theorem 3.6.1 of [20], then the long-run average cost, denoted by $C(M)$, is given as follows:

$$C(M) = \frac{c(0, M)}{E[T_M]}.$$

For $y \in R_{+}$, define $t(y) = 1 - Q(y, R_{+}, \infty)$. Using Corollaries 2.22 (i), 2.26 and 2.28, it follows that

$$C(M) = \frac{c_1 + \int_{[x,M]} R(0, dy, \infty)(kt(y) + m(y)c(y))}{\int_{[0,M]} R(0, dy, \infty)m(y)}. \tag{2.16}$$

The proof of the following theorem follows from (2.16) in a manner similar to the proof of Theorem 2.30, and hence is omitted.

Theorem 2.32 Assume that $E[T_M] < \infty$, $A \to R(0, A, \infty)$ is absolutely continuous with respect to the Lebesgue measure on R_{+}, and

(i) $y \to m(y)$ is decreasing,
(ii) $y \to Q(y, R_{+}, \infty)$ is decreasing,
(iii) $y \to c(y)$ is increasing and bounded.

Then the optimal value M^* is the unique solution of the integral equation

$$c_1 = \int_{[0,M]} R(0, dy, \infty)\{k(\frac{m(y)}{m(M)}(t(M) - t(y)) + m(y)(c(M) - c(y))\}. \tag{2.17}$$

If no solution exists, then the optimal policy is to wait until failure for replacement.

2.6 Multi-Level Control-Limit Maintenance Policies for Non-Continuously Monitored Degradable Systems

Assume that a system is subject to a subordinator degradation process, denoted by Z. The device has a nominal distribution Y, and it fails once the degradation exceeds level Y. The sates of the system are only observed by inspection, and failures are only detected by inspection. For $n \geq 1$, let $\overset{\wedge}{\tau}_n$ be the time of the nth inspection, $\overset{\wedge}{\tau}_0 = 0$. For $n \geq 0$, let $\hat{X}_n = Z_{\overset{\wedge}{\tau}_n}$ be the degradation level at the time of the nth inspection. There are $N (N > 1)$ threshold values $\{l_0, l_1, .., l_{N-1}\}$, such that $0 = l_0 < l_1 < l_2 < ...l_{N-1}$, and define $l_N = \infty$. For $n \geq 0$, we let $T_n = nT$, where T is a positive number. For $x \in R_{+}$, suppose that $D(x) = \inf\{k : l_k \leq x < l_{k+1}\}$. For $n \geq 0$, the sequence of scheduled inspection times is defined as follows: Given $\hat{X}_n, \overset{\wedge}{\tau}_n$, the time of the next inspection $\overset{\wedge}{\tau}_{n+1}$ is equal to $\overset{\wedge}{\tau}_n + T_{N-D(\hat{X}_n)}$. Assume that the preventive maintenance level is equal to M, and $M > l_{N-1}$.

The following maintenance scheme is adopted: For $n \geq 0$, and given $\hat{\tau}_n, \hat{X}_n$

(1) If $\hat{X}_n < M \wedge Y$, then no maintenance is performed, and next inspection is scheduled at time $\hat{\tau}_{n+1}$;

(2) If $M \leq \hat{X}_n < Y$, a preventive maintenance is performed at a cost $c_1 > 0$;

(3) If $\hat{X}_n \geq Y$, a corrective maintenance is performed at a cost $c_1 + k, k > 0$;

(4) If at inspection, the degradation level is y, a penalty cost is accrued at a rate $c(y)$ until the next inspection occurs;

(5) If $M \leq \hat{X}_n < Y$ or $\hat{X}_n \geq Y$, then the time of the next scheduled inspection is equal to $\hat{\tau}_n + T_N$.

We note that (5) above simply restates the assumption that after each maintenance, the system is as good as new with zero degradation.

Theorem 2.33 The the process $\{(\hat{X}_n, \hat{\tau}_n), n \geq 0\}$ is a a homogeneous Markov renewal process.

Proof For $n \geq 0$, for any Borel set $A \subset R_+$, and $t \in R_+$

$$\begin{aligned} P\{\hat{X}_{n+1} \in A, \hat{\tau}_{n+1} - \hat{\tau}_n \leq t \mid \hat{X}_n = x, \hat{\tau}_n = s\} \\ = P\{Z_{\hat{\tau}_{n+1}} \in A, \hat{\tau}_{n+1} - \hat{\tau}_n \leq t \mid \hat{X}_n = x, \hat{\tau}_n\} \\ = P\{Z_{T_{N-D(x)}} \in A - x, T_{N-D(x)} \leq t\}, \end{aligned}$$

where the last equation follows because the process Z has stationary independent increments, and the definition of $\hat{\tau}_{n+1}$. Our assertion is proven, since the term in the last equation above does not depend on the value of $\hat{\tau}_n$, and is the same for all n. ■

Let $\hat{Q}(x, A, t)$, $\hat{R}(x, A, t)$ be the semi-Markov and Markov renewal kernels associated with the above process, respectively. Suppose that, $\hat{Y}$, is the corresponding semi-Markov process. The time of first maintenance (T_M) is as follows:

$$T_M = \inf\{t : \hat{Y}_t \geq M \wedge Y\}$$

As in Sect. 2.5 we let

$$L = \inf\{n : \hat{\tau}_n = T_M\}.$$

Suppose, for $t \in R_+, n \geq 0$,

$$X_n = \{\hat{X}_n, n < L\},$$

and

$$\tau_n = \{\overset{\wedge}{\tau}_n, n \leq L\}.$$

It is clear that the Markov renewal process $\{(X_n \tau_n), n \geq 0\}$ has the interval $[0, M]$ as its state space. Let $Q(x, A, t)$, $R(x, A, t)$ be the semi-Markov and Markov renewal kernels associated with this process, respectively. Following similar steps to the ones used in proving Lemma 2.19. and Proposition 2.20, it can be shown that, for any for $t \in R_+$, and Borel set $A \subseteq [0, M]$

$$Q(x, A, t) = \int_A \hat{Q}(x, dy, t) \frac{\bar{G}(y)}{\bar{G}(x)},$$

and

$$R(x, A, t) = \int_A \hat{R}(x, dy, t) \frac{\bar{G}(y)}{\bar{G}(x)}.$$

For $y, \alpha \in R_+$, let

$$h_\alpha(y) = \frac{1}{\alpha} E(1 - e^{-\alpha T_{N-D(y)}}).$$

Note that this function is decreasing in y; furthermore $E[T_{N-D(y)}]$ is decreasing in y. The proof of the following theorem follows in a manner similar to the proof of Theorem 2.30, we omit the proof.

Theorem 2.34 Assume that $A \to R_\alpha(0, A)$ is absolutely continuous with respect to the Lebesgue measure on R_+, and

(i) $y \to [Q_\alpha(y, R_+) - Q_\alpha(y, R_+)]$ is decreasing,
(ii) $y \to c(y)$ is increasing and bounded.

Then the optimal value M^* is the unique solution of the integral equation (2.15). If no solution exists, then the optimal control-limit policy is to wait until failure for replacement.

The following theorem is analogous to Theorem 2.32, its proof is very much similar and is omitted.

Theorem 2.35 Assume that $E[T_M\} < \infty$, $A \to R(0, A, \infty)$ is absolutely continuous with respect to the Lebesgue measure on R_+, and

(i) $y \to Q(y, R_+, \infty)]$ is increasing,
(ii) $y \to c(y)$ is increasing and bounded.

Then the optimal value M^* is the unique solution of the integral equation (2.16). If no solution exists, then the optimal control-limit policy is to wait until failure for replacement.

We finish this section by remarking that the results therein streamline and extend those obtained in [6].

2.7 Examples

In this section we give applications of the results obtained in Sects. 2.2–2.6. Particular applications are considered where the degradation process is a nonhomogeneous gamma, inverse Gaussian, and compound Poisson process. In each case we discuss properties of the life distribution and the behavior of the failure rate as well as the optimal maintenance and replacement policy.

Example 1 Assume that a device is subject to a nonhomogeneous subordinator degradation process, and the degradation is monitored continuously. Suppose that the cost of a replacement before failure $c_1(.) = c - k, 0 < k < c$, on B, and equal to c on $\bar{B}$, where c is the cost of a replacement at failure. Then, $c_2(.) \equiv c - c_1\,(.) = k$ on B, and is equal to 0 on $\bar{B}$. Assume that, on B, $\bar{G}$ is absolutely continuous on and have bounded increasing failure rate. Then, the corresponding infinitesimal generator is of the form

$$Ac_2(x) = k \int\limits_B [\bar{G}(x+y)/\bar{G}(x)\} - 1]\upsilon(dy),$$

which is clearly decreasing on B. From Theorem 2.17, it follows that the control-limit policy is the optimal policy.

Example 2 Suppose that a device is subject to a nonhomogeneous subordinator degradation process (X), and that the degradation is monitored continuously. Assume the same cost structure in Example 1, and the nominal life variable Y has exponential distribution, i.e., $\bar{G}(x) = \exp(-\theta x), \theta > 0$. Assume that the potential ($\overset{\wedge}{\mathbf{R}}$) of the corresponding stationary subordinator ($\hat{X}$) is absolutely continuous with respect to the Lebesgue measure. Then the optimal maintenance policy is to wait till failure for replacement. This is follows, since (for the corresponding stationary subordinator) from Theorem 2.18, the long-run average cost associated with any control-limit policy is

$$\begin{aligned}\psi(x) &= \frac{c - E(c^*(\hat{X}\tau_x))}{\mathbf{R}(x)} \\ &= \frac{c - kE(\bar{G}(\hat{X}\tau_x))}{\mathbf{R}(x)} \\ &= \frac{c - kE(\exp(-\theta\hat{X}\tau_x))}{\mathbf{R}(x)}.\end{aligned}$$

From Lemma 3.10 (of Chap. 3), it follows that

$$E(\exp(-\theta \hat{X}\tau_x)) = \phi(\theta) \int_x^{\infty} e^{-\theta y} \overset{\wedge}{\mathbf{R}}(dy),$$

where $\phi(\theta)$ is the Laplace exponent of the process $\hat{X}$ and, as defined in Theorem 2.18,

$$\mathbf{R}(x) = \int_0^x e^{-\theta y} \overset{\wedge}{\mathbf{R}}(dy).$$

Since, $\int_0^{\infty} e^{-\theta y} \overset{\wedge}{\mathbf{R}}(dy) = 1/\phi(\theta)$,

$$\psi(0) = \infty, \quad \text{and} \quad \psi(\infty) = c\phi(\theta).$$

Direct differentiation of $\psi(x)$, and using the above formulas for $E(\exp(-\theta \hat{X}\tau_x))$, $\mathbf{R}(x)$, it is easy to see that

$$\psi'(x) \overset{sign}{=} k - c.$$

Since, by assumptions, $k < c$, then $\psi(x)$ is decreasing in x, and thus the optimal policy is to wait till failure for replacement.

Example 3 Assume that the degradation process is a nonhomogeneous Poisson process, denoted by X. Moreover, if the jump distribution density is PF_2, then as shown In Example 3 of Sect. 2.2, the probability transition function of the corresponding homogeneous compound Poisson process is TP_2, thus the conclusions in Theorems 2.9 and 2.10 hold. In particular, with $\bar{F}$ as given in (2.6), we have: (i) If $\bar{G}$ is IFR and Λ is convex, then $\bar{F}$ is *IFR*; ii) If $\bar{G}$ is IFRA and Λ is starshaped, then $\bar{F}$ is IFRA.

Suppose that the degradation is observed continuously, and assume the cost structure given in Example 1. The long-run average cost, corresponding to the optimal control limit policy is computed using Theorem 2.18, where from (1.34), the potential associated with the corresponding homogeneous process ($\hat{X}$) is given as follows:

$$\overset{\wedge}{\mathbf{R}}(x) = \frac{1}{\lambda} M(x),$$

where $M(x)$ is the renewal function corresponding to the distribution function of the jump sizes, distribution of $\hat{X}\tau_x$ is given in (3.63), of Chap. 3, and $c^*(x) = k\bar{G}(x)$.

Example 4 Assume that the degradation process is a nonhomogeneous inverse Gaussian process, denoted by Z. In Example 2 of Sect. 2.2, it is shown that the probability transition function of the corresponding homogeneous inverse Gaussian process is TP_2. Thus the life distribution of the device belongs to the different classes given in Theorems 2.9 and 2.10, under the appropriate assumptions on $\bar{G}$ and Λ, given therein.

Suppose that the degradation is observed continuously, and assume the same cost structure given in Example 1. In this case, the control limit policy is the optimal policy, and the corresponding the long-run average cost can be computed using Theorem 2.19, where from (1.35) the potential associated with the corresponding homogeneous process ($\hat{Z}$) is absolutely continuous with density

$$r(y) = \frac{\sigma}{\sqrt{y}}\varphi(\sqrt{y}\mu/\sigma) + (\frac{\mu}{2}) \ \text{erf} \ c(\sqrt{y}\frac{\mu\sqrt{y}}{\sqrt{2\sigma^2}}),$$

where φ (.) is the standard normal density, furthermore the distribution of $\hat{Z}_{\tau_x}$ is given in (3.70), of Chap. 3.

Example 5 Suppose that a device is subject to degradation, and the degradation process Z is a subordinator. Assume that the degradation process is maintained periodically, using the one-level control-limit maintenance policy described in Sect. 2.5. Assume that the nominal life distribution is IFR with right-tail probability $\bar{G}$, then $t(y) \overset{def}{=} 1 - Q(x, R_+, \infty) = 1 - \frac{E(\bar{G}(X_1+y))}{\bar{G}(y)}$, is increasing in y. Assume that, for all x, $W(y) = W$, where W is a positive random variable. In this case, the values of the deterioration at inspection times (if no maintenance to be performed), $\{\hat{X}_n, n \geq 1\}$, form a renewal process and $\hat{X}_1 = Z_W$, and we denote its renewal function by $\hat{R}$. Let $m = E(W)$, and assume that the cost function $c(x)$ is increasing, then it follows from Theorem 2.32 that the optimal preventive maintenance level (using the long-run average criterion) is the solution of the integral equation

$$c_1 = \int_{[0,M]} \hat{R}(dy)\bar{G}(y)\{k(t(M) - t(y)) + m(c(M) - c(y))\}.$$

Example 6 Let $C_M = c(M)R(0, [0, M], \infty) - (c_1/m)$. Assume that the conditions given in Example 5 are satisfied, and the nominal life is either exponential or degenerate at ∞. In both cases, $t(y)$ is the constant function. In this case, the optimal preventive maintenance level is the solution of the integral equation

$$C_M = \int_{[0,M]} \hat{R}(dy)\bar{G}(y)c(y)$$

If $c(y)$ is constant, then the above equation has no solution and the optimal preventive maintenance level $M^* = \infty$.

Example 7 Assume that a device is subject to a nonhomogeneous gamma degradation process, with scale parameter β. Since (as shown in Example 1 of Sect. 2.2) the transition function of the stationary gamma process is TP_2, then Theorems 2.9 and 2.10 apply. Let $\bar{F}$ is given in (2.6), then in particular, we have: (i) If $\bar{G}$ is IFR and Λ is convex, then $\bar{F}$ is IFR. (ii) If $\bar{G}$ is IFRA and Λ is starshaped, then $\bar{F}$ is IFRA.

Now assume that the one-level control limit policy, is the maintenance policy used for maintaining the device. Suppose that the system is periodically inspected at times $1, 2, 3, \ldots$, that is for all $x \in R_+$, the random variable $W(x)$ is equal to a 1 almost surely. It follows that the renewal function $(\hat{R}(x))$ in Example 5 above is equal to $1 + \beta x$. If $\bar{G}(x) = 1, x \in R_+$, $c(x) = \frac{x}{x+a}$, $a > 0$, and $D(M) = \ln(\frac{M+a}{a}) - a(\frac{M}{M=a})$, then it is seen (after some straightforward calculations, which we omit) that the optimal preventive maintenance level (M^*) is the solution of the equation

$$D(M) = c_1/\beta.$$

If $\bar{G}(x) = \mathbf{I}_{\{x<L\}}$, $L > 0$, $c(x) \equiv c > 0$, then it follows that the optimal preventive maintenance level (M^*) is the solution of the equation

$$D_1(M) = (c_1/k)e^{\beta L} - 1,$$

where $D_1(M) = (\beta M - 1)e^{\beta M}$.

2.8 Inference for the Parameters of the Degradation Process

Throughout we assume that times $0 = t_0 < t_1 < \cdots < t_n$ are fixed, and for $1 \le i \le n$, we define $w_i = \Lambda(t_i) - \Lambda(t_{i-1})$. The following lemma follows easily from the fact that the process given is a Levy process, we omit the proof.

Lemma 2.36 Let $X = (X_t, t \ge 0)$ be a nonhomogeneous Levy process (in the sense described in Sect. 1.8), with transition probability $p(t, x)$. Suppose that $f(t, x)$ is the transition function of the stationary Levy process $(X_{\Lambda^{-1}(t)}, t \ge 0)$. For $\underline{x} = (x_0, x_1, \ldots, x_n) \in R^{n+1}$, and for $1 \le i \le n$, we define $z_i = x_i - x_{i-1}$. Suppose that $\underline{\alpha}$ is the vector of parameters of the process X. Given the observations $\underline{z} = (z_1, \ldots, z_n)$, the likelihood equation is as follows:

$$\pounds(\underline{\alpha}, \underline{z}) = \prod_{i=1}^{n} f(\underline{\alpha}, w_i, z_i).$$

Many authors (see [24, 25] and the references therein), used the Brownian motion with positive drift to model degradation over time. The advantage of this model is that the increments have normal distribution and thus the estimation of the parameters of the degradation process is rather straightforward. A disadvantage of this process is that is not increasing, takes negative as well as positive values. However, if the mean rate is large and variance is small, the chance that the increments are negative become small. We will not discuss this model here, instead we will concentrate on the cases where the degradation process is either a nonhomogeneous gamma or inverse Gaussian process.

We start first with the case where degradation process is a nonhomogeneous gamma process. The following theorem gives the maximum likelihood estimators of the parameters of this process.

Theorem 2.37 Suppose that the degradation process X is a nonhomogeneous gamma process (in the sense described in Sects. 1.4 and 1.8) Let $\underline{x}$, $(z_1, ..., z_n)$ be the observations defined in Lemma 2.36, and define $\gamma_n = \sum_{i=1}^{n} z_i$. The maximum likelihood estimator $\hat{\alpha}$, of the parameter α, is the numerical solution of the equation

$$\sum_{i=1}^{n} w_i(\Psi(\alpha w_i) - \log(z_i)) = \Lambda(t_n) \log(\alpha\Lambda(t_n)/\gamma_n),$$

where Ψ is the Digamma function, $\Psi(x) = \frac{\Gamma'(x)}{\Gamma(x)}$, and maximum likelihood estimator of the parameter β is

$$\hat{\beta} = \frac{\hat{\alpha}\Lambda(t_n)}{\gamma_n}.$$

Proof Given the observations vector $\underline{z} = (z_1, ..., x_n)$, the likelihood function is as follows:

$$\mathcal{L}(\alpha, \beta, \underline{z}) = \prod_{i=1}^{n} e^{-z_i\beta} \frac{z_i^{\alpha w_i - 1}}{\Gamma(\alpha w_i)} (\beta)^{\alpha w_i}$$

Thus, the log likelihood functions is of the form

$$l(\alpha, \beta, \underline{z}) = \sum_{i=1}^{n}(\alpha w_i - 1)\log(z_i) - \beta\sum_{i=1}^{n} z_i + \alpha\log(\beta)\sum_{i=1}^{n} w_i - \sum_{i=1}^{n}\log(\Gamma(\alpha w_i)). \tag{2.18}$$

Differentiating (2.18) with respect to β and setting the result equal to zero we have

$$\sum_{i=1}^{n} z_i - \frac{\alpha}{\beta}\sum_{i=1}^{n} w_i = 0. \tag{2.19}$$

Since, $\sum_{i=1}^{n} z_i = \delta_n$ and $\sum_{i=1}^{n} u_i = \Lambda(t_n)$, it follows from (2.19) that

$$\hat{\beta} = \frac{\hat{\alpha}\Lambda(t_n)}{\delta_n}, \tag{2.20}$$

Differentiating (2.18) with respect to α, setting the result equal to zero, we have

$$\sum_{i=1}^{n} w_i \log(z_i) + \Lambda(t_n)\log(\beta) - \sum_{i=1}^{n} \Psi(\alpha w_i) w_i = 0 \tag{2.21}$$

From (2.20) and (2.21), it follows that the maximum likelihood estimator of the parameter α is the solution to the equation

$$\sum_{i=1}^{n} w_i(\Psi(\alpha w_i) - \log(z_i)) = \Lambda(t_n)\log(\alpha\Lambda(t_n)/\sum_{i=1}^{n} z_i).$$

and our assertion is thus proven. ■

The following theorem gives the empirical estimators of the parameters α and β.

Theorem 2.38 Suppose that the degradation process X is a nonhomogeneous gamma process. Let $\underline{z}$ and γ_n be as given in Theorem 2.37, define $\delta_n = \frac{\gamma_n}{\Lambda(t_n)}$, then the empirical estimates of α and β (denoted by α_E and β_E, respectively) are as follows:

$$\beta_E = \frac{\gamma_n(1 - \{\sum_{i=1}^{n} w_i^2/[\sum_{i=1}^{n} w_i]^2\})}{\sum_{i=1}^{n} (z_i - w_i\delta_n)^2},$$

and

$$\alpha_E = \frac{\gamma_n^2(1 - \{\sum_{i=1}^{n} w_i^2/[\sum_{i=1}^{n} w_i]^2\})}{\Lambda(t_n)\sum_{i=1}^{n} (z_i - w_i\delta_n)^2}$$

Proof For $0 \leq i \leq n$, we abuse the notation and denote X_{t_i} by X_i, for $1 \leq i \leq n$, we define $Z_i = X_i - X_{i-1}$, and we let $U_n = \sum_{i=1}^{n} Z_i$. Note that $(z_1, \ldots, z_n)$ are the observed values of the random variables $(Z_1, \ldots, Z_n)$, and γ_n is the observed value of U_n. Denote $\Lambda(t_n)$ by b_n, and define the random variable

$$T_n = \frac{U_n}{b_n}.$$

Then,

$$E(T_n) = \frac{\alpha b_n}{\beta b_n} = \alpha/\beta.$$

Thus the empirical estimators of α and β satisfy the following

$$\alpha_E/\beta_E = T_n. \tag{2.22}$$

Define

$$Y_n = \frac{b_n \sum_{i=1}^{n} (Z_i - w_i T_n)^2}{(b_n^2 - \sum_{i=1}^{n} w_i^2)}. \tag{2.23}$$

We have

$$\begin{aligned} E\sum_{i=1}^{n}(Z_i - w_i T_n)^2 &= E\sum_{i=1}^{n}(Z_i - \alpha w_i/\beta - w_i T_n + \alpha w_i/\beta)^2 \\ &= E\sum_{i=1}^{n}(Z_i - \alpha w_i/\beta)^2 + E\sum_{i=1}^{n} w_i^2 (T_n - \alpha/\beta)^2 \\ &\quad - 2E\sum_{i=1}^{n} w_i (Z_i - \alpha w_i/\beta)(T_n - \alpha/\beta). \end{aligned} \tag{2.24}$$

To evaluate the first term in (2.24) we write

$$\begin{aligned} E\sum_{i=1}^{n}(Z_i - \alpha w_i/\beta)^2 &= \sum_{i=1}^{n} var(Z_i) \\ &= \alpha \sum_{i=1}^{n} w_i/\beta^2 \\ &= \alpha b_n/\beta^2. \end{aligned} \tag{2.25}$$

The second term in (2.24) is computed as follows:

$$E\sum_{i=1}^{n} w_i^2 (T_n - \alpha/\beta)^2 = \sum_{i=1}^{n} w_i^2 var(T_n)$$

$$= var(T_n) \sum_{i=1}^{n} w_i^2$$

$$= (\alpha/b_n\beta^2) \sum_{i=1}^{n} w_i^2. \tag{2.26}$$

To compute the third term in (2.24) we write

$$E \sum_{i=1}^{n} w_i(Z_i - \alpha w_i/\beta)(T_n - \alpha/\beta) = \frac{1}{b_n} E \sum_{j=1}^{n}\sum_{i=1}^{n} w_i(Z_i - \alpha w_i/\beta)(Z_j - \alpha w_i/\beta)$$

$$= \frac{1}{b_n} \sum_{j=1}^{n}\sum_{i=1}^{n} w_i cov(Z_i, Z_j)$$

$$= \frac{1}{b_n} \sum_{i=1}^{n} w_i var(Z_i)$$

$$= (\alpha/b_n\beta^2) \sum_{i=1}^{n} w_i^2, \tag{2.27}$$

where the third term above follows from the second term, since the random variables $Z_1, Z_2, \ldots$ are independent.

Using (2.23)–(2.27), it follows that

$$E(Y_n) = \alpha/\beta^2.$$

Thus the empirical estimators of α and β satisfy

$$\alpha_E/\beta_E^2 = y_n. \tag{2.28}$$

From (2.22) and (2.28), it follows that

$$\beta_E = \frac{\sum_{i=1}^{n} z_i}{b_n y_n}, \tag{2.29}$$

and

$$\alpha_E = \frac{\gamma_n}{\beta_E b_n}$$

$$= \frac{(\gamma_n)^2}{b_n^2 y_n}. \tag{2.30}$$

Our assertion is proven using (2.23), (2.29) and (2.30). ■

The following gives the Bayes estimate of the parameter β assuming α is known.

Theorem 2.39 Assume that the degradation process X is a nonhomogeneous gamma process. Let $\underline{z}$ and γ_n be as given in Theorem 2.37, and suppose that α is known. Assume that the prior distribution of the random variable β is a gamma distribution with scale and shape parameters a and b respectively, and denote its density by $g(\beta)$. Then, the Bayes estimate of the parameter

$$\overset{\approx}{\beta} = \frac{\alpha\Lambda(t_n) + a + 1}{\gamma_n + b}.$$

Proof Let $\mathcal{L}(\alpha, \beta, \underline{z})$ be the likelihood function given in the proof of Theorem 2.37. Note that the posterior density of the parameter β, given α and $\underline{z}$ is of the form

$$f(\beta \mid \alpha, \underline{z}) = \frac{\mathcal{L}(\alpha, \beta, \underline{z})g(\beta)}{\int \mathcal{L}(\alpha, y, \underline{z})g(y)dy},$$

and the Bayes estimate of β is the expected value of the random variable whose density is $f(\beta \mid \alpha, \underline{z})$.

Note that

$$\begin{aligned}\mathcal{L}(\alpha, \beta, \underline{x}) &= \prod_{i=1}^{n} e^{-z_i\beta} \frac{z_i^{\alpha w_i - 1}}{\Gamma(\alpha w_i)} (\beta)^{\alpha w_i} \\ &= e^{-\gamma_n\beta}(\beta)^{\alpha\Lambda(t_n)} \prod_{i=1}^{n} \frac{z_i^{\alpha w_i - 1}}{\Gamma(\alpha w_i)}. \qquad (2.31)\end{aligned}$$

Note that the first two terms in (2.31) multiplied by $\frac{(\gamma_n)^{\alpha\Lambda(t_n)+1}}{\Gamma(\alpha\Lambda(t_{ni})+1)}$ gives us the density function of a gamma random variable with with scale and shape parameters γ_n and $\alpha\Lambda(t_n)+1$, respectively. It is known that the posterior density of this gamma random variable when the prior is a gamma distribution with scale and shape parameters a and b, respectively, is a gamma random variable with scale and shape parameters $\gamma_n + a$ and $\alpha\Lambda(t_n) + b + 1$ (see p. 183 of [38]). Our assertion follows since the mean of a gamma random variable with scale and shape parameters $\gamma_n + a$ and $\alpha\Lambda(t_n) + b + 1$ is equal to $\frac{\alpha\Lambda(t_n)+b+1}{\gamma_n+a}$. ■

We note that the results in Theorems 2.37 and 2.38, generalize the results obtained in [13], where it is assumed that $\Lambda(t) = t^p$.

The following theorem gives the maximum likelihood estimators of the parameters of the degradation process, when the latter is an inverse Brownian motion.

Theorem 2.40 Suppose that the degradation process X is a nonhomogeneous inverse Brownian motion with parameters μ and σ^2 as given in (1.5). Let $\underline{z}$ and γ_n be as given in Theorem 2.37, then the maximum likelihood estimator of the parameter μ and σ^2 are as follows:

$$\hat{\mu} = \frac{\Lambda(t_n)}{\gamma_n}, \tag{2.32}$$

and

$$\hat{\sigma}^2 = \frac{1}{n}\sum_{i=1}^{n} w_i\left(\frac{w_i}{z_i} - \hat{\mu}\right). \tag{2.33}$$

Proof From (1.5), it follows that the likelihood equation is of the form

$$\mathcal{L}(\alpha, \beta, \underline{x}) = (\sigma^2)^{-n/2} \prod_{i=1}^{n} \frac{w_i}{\sqrt{2\pi z_i^3}} e^{-\frac{(\mu z_i - w_i)^2}{2z_i\sigma^2}}.$$

Thus the log likelihood function is equal to

$$l(\alpha, \beta, \underline{x}) = c + (-n/2)\ln(\sigma^2) - \sum_{i=1}^{n} \frac{(\mu z_i - w_i)^2}{2z_i\sigma^2}, \tag{2.34}$$

where c is a function of $\underline{w}$, $\underline{z}$ only, which is easily determined.

Differentiating the right hand side of (2.34) with respect to μ and equating the result with zero we have

$$\sum_{i=1}^{n} (\mu z_i - w_i) = 0, \tag{2.35}$$

the corresponding equation when we differentiate (2.34) with respect to σ^2 is

$$\sum_{i=1}^{n} \frac{(\mu z_i - w_i)^2}{z_i\sigma^4} - \frac{n}{\sigma^2} = 0. \tag{2.36}$$

We note that (2.32) is immediate from (2.35); from (2.36), we have

$$\hat{\sigma}^2 = \frac{1}{n}\sum_{i=1}^{n} \frac{(\hat{\mu} z_i - w_i)^2}{z_i}. \tag{2.37}$$

Our assertion follows immediately from (2.37), since

$$\frac{1}{n}\sum_{i=1}^{n}\frac{(\hat{\mu}z_i - w_i)^2}{z_i} = \frac{1}{n}\sum_{i=1}^{n} w_i(\frac{w_i}{z_i} - \hat{\mu}),$$

which immediately follows from (2.35), after straightforward manipulations, which we omit. ∎

The following theorem shows that $\hat{\mu}$ and $\hat{\sigma}^2$ given in above are sufficient estimators of μ and σ^2, respectively.

Theorem 2.41 Suppose that the degradation process X is a nonhomogeneous inverse Brownian motion with parameters μ and σ^2. Let $\underline{z}$ and γ_n be as given in Theorem 2.37, then $\hat{\mu}$ and $\hat{\sigma}^2$ given in (2.32) and (2.33), are sufficient estimators of the parameters μ and σ^2, respectively.

Proof We write

$$\begin{aligned}\mathcal{L}(\alpha, \beta, \underline{z}) &= (\sigma^2)^{-n/2}\prod_{i=1}^{n}\frac{w_i}{\sqrt{2\pi z_i^3}}e^{-\frac{(\mu z_i - w_i)^2}{2z_i\sigma^2}} \\ &= (\prod_{i=1}^{n}\frac{w_i}{\sqrt{2\pi z_i^3}})(\sigma^2)^{-n/2}\exp\{-\sum_{i=1}^{n}\frac{(\mu z_i - w_i)^2}{2z_i\sigma^2}\} \\ &= (\prod_{i=1}^{n}\frac{w_i}{\sqrt{2\pi z_i^3}})[(\sigma^2)^{-n/2}\exp\{-\frac{n\hat{\sigma}^2}{2\sigma^2} - \frac{\Lambda(t_n)(\mu - \hat{\mu})^2}{2\hat{\mu}\sigma^2}\}], \end{aligned} \tag{2.38}$$

where the last equality follows from (2.32), (2.33) and (2.35).

Our assertion follows from the factorization theorem, since the first term in (2.38) depends only on the observation vector $\underline{z}$, and the second term depends on the parameters and the estimators, and not the observation vector. ∎

We now turn our attention to the case where the observed degradation is subject to error. Let $X = (X_t, t \geq 0)$ be a nonhomogeneous Levy process, and transition probability $p(t, x)$. Suppose that $f(t, x)$ is the transition function of the corresponding stationary Levy process.

For $t \in R_+$, we define

$$Y_t = X_t + \epsilon_t,$$

where ϵ_t is independent from X_t, and has as a normal distribution with mean zero and variance σ^2.

Assume that the process Y is observed at times $0 = t_0 < t_1 < \cdots < t_n$. For $\underline{y} = (y_0, y_1, \ldots, n_n) \in R^{n+1}$, and for $1 \leq i \leq n$, we define $z_i = y_i - y_{i-1}$, we abuse the notation and write ϵ_i for $\epsilon_{\Lambda(t_i)}$, and $V_i = \epsilon_i - \epsilon_{i-1}$. It follows that the vector $\underline{V} = (V_1, ..., V_n)$ has a multivariate normal distribution with zero mean and covariance matrix whose i, jth value $\sigma_{i,j} = -\sigma^2$, if $j - i = \pm 1, = 2\sigma^2$, if $i = j$ and zero otherwise. Suppose that $\underline{\alpha}$ is the vector of parameters of the process X. Given the observations $\underline{z} = (z_1, ..., z_n)$, the likelihood equation is as follows:

$$£(\underline{\alpha}, \underline{z}) = E[\prod_{i=1}^{n} f(\underline{\alpha}, w_i, z_i - V_i)], \tag{2.39}$$

where the expectation is taken with respect to the variables $V_1, ..., V_n$.

To compute the maximum likelihood estimator of the vector $\underline{\alpha}$, we have two possible procedures.

Procedure 1 Given $\underline{V} = \underline{v}$, compute the MLE $\hat{\underline{\alpha}}(\underline{z}, \underline{v})$. Then the maximum likelihood estimator of $\underline{\alpha}$ is equal to $E[\hat{\underline{\alpha}}(\underline{z}, \underline{V})]$.

Procedure 2 Simulate N values of the vector $\underline{\epsilon} = (\epsilon_1, ...\epsilon_n)$, and $1 \le k \le N$, denote the kth simulated vector by $\underline{\epsilon}^{(k)} = (\epsilon_1^{(k)}, ..., \epsilon_n^{(k)})$. For $1 \le k \le N$, $1 \le i \le n$, let $v_i^{(k)} = \epsilon_i^{(k)} - \epsilon_{i-1}^{(k)}$. For large N, the likelihood equation in (2.39) can be approximated by

$$\frac{1}{N}\sum_{k=1}^{N}\prod_{i=1}^{n} f(\underline{\alpha}, w_i, z_i - v_i^{(k)}).$$

From the last equation an approximation of the maximum likelihood estimator of $\underline{\alpha}$ can be obtained, using standard methods.

For special types of subordinator deterioration processes, the above model has been examined (See [17], and the reference therein).

References

1. Abdel-Hameed M (1975) A gamma wear process. IEEE Trans Reliab 24(2):152–153
2. Abdel-Hameed M (1977) Optimal replacement policies for devices subject to a gamma wear process. The theory and applications of reliability; with emphasis on Bayesian and nonparametric methods, Academic Press, New York, pp 397–412
3. van Noortwijk J. M. (1998) Optimal replacement decisions for structures under stochastic deterioration. Proceedings of the eighth IFIP WG 7.5 working conference on reliability and optimization of structural systems, 273–280. University of Michigan: Ann Arbor.
4. Abdel-Hameed M (1987) Inspection and maintenance policies of devices subject to deterioration. Adv Appl Prob 19:917–931
5. Castanier B, Berenguer C, Grall A (2003) A sequential condition-based repair/replacement policy with non-periodic inspections for a system subject to continuous wear. Appl Stochastic Models Bus Ind 19:327–347

6. Grall A, Berenguer C, Dieulle L (2002) A condition-based maintenance policy for stochastically deteriorating systems. Reliab Eng Syst Saf 76:167–180
7. Grall A, Dieulle L, Berenguer C, Roussignol M (2002) Continuous-time predictive-maintenance scheduling for a deteriorating system. IEEE Trans Reliab 51:141–150
8. Madanat S, Mishalani R (1995) Estimation of infrastructure transition probabilities from condition rating data. J Infrastruct Syst 1(2):120–125
9. Mori Y, Ellingwood BR (1994) Maintaining: reliability of concrete structures, I: role of inspection/repair. J Struct Eng 120:45–824
10. Newby M, Dagg R (2002) Optimal inspection and maintenance for stochastically deteriorating systems I: average cost criteria. J Indian Stat Assoc 40:98–169
11. Newby M, Dagg R (2003) Optimal inspection and maintenance for stochastically deteriorating systems II: discounted cost criterion. J Indian Statist Assoc 41:9–27
12. Newby M, Dagg R (2003) Inspection and maintenance for stochastically deteriorating systems. World Scientific, Mathematical and statistical methods in reliability
13. Cinlar E, Bazant ZP, Osman E (1977) Stochastic process for extrapolating concrete creep. J Eng Mech Div 103:1069–1088
14. Lawless J, Crowder M (2004) Covariates and random effects in a gamma process model with application to degradation and failure. Lifetime Data Anal 10:27–213
15. Frangopol DM, Kallen MJ, van Noortwijk JM (2004) Probabilistic models for life-cycle performance of deteriorating structures: review and future directions. Prog Struct Eng Mater 6:197–212
16. Speijker UP, van Noortwijk JM, Kok M, Cooke RM (2000) Optimal maintenance decisions for dikes. Probab Eng Inf Sci 14:21–101
17. Kallen MJ, van Noortwijk JM (2005) Optimal maintenance decisions under imperfect inspection. Reliab Eng Syst Saf 90:177–185
18. Barlow RE, Proschan F (1975) Statistical theory of reliability and life testing. Holt, Reinhart and Winston, NewYok
19. Karlin S (1968) Total Positivity, vol 0049. Stanford University Press, London
20. Ross SM (1983) Stochastic processes. Wiley, New York
21. Oksendal B (2000) Stochastic differential equations. Springer, Berlin
22. Liao H, Elsayed EA, Chan LY (2006) Maintenance of continuously monitored degrading systems. Eur J Oper Res 175:35–821
23. Cinlar E (1975) Introduction to stochastic processes. Prentice Hall, New Jersey
24. Park C, Padgett WJ (2006) Stochastic degradation models with several accelerating variables. IEEE Trans Reliab 55:379–390
25. Whitmore GA, Schenkelberg F (1997) Modelling accelerated degradation data using Wiener diffusion with a time scale transformation. Lifetime Data Anal 3:27–45
26. Abdel-Hameed M (1984) Life distribution properties of devices subject to a pure jump damage process. J Appl Prob 21:816–825
27. Abdel-Hameed MS (2004) Optimal predictive maintenance policies for a deteriorating system: the total discounted cost and the long-run average cost cases. Commun Stat 33:735–745
28. Bakker JD, van Noortwijk JM (2004) Inspection validation model for lifecycle analysis. In: Proceedings of the second international conference on bridge maintenance, safety and management and cost Kyoto, Taylor and Francis, Japan, pp 18–22
29. Barlow RE, Proschan F (1965) Mathematical theory of reliability, Wiley, NewYok
30. Berenguer C, Dieulle L, Grall A, Roussignol M (2001) Two monitoring maintenance strategies for a stochastically deteriorating system. In: Proceedings of the 10th international symposium on applied stochastic models and data analysis, Compiegne, France, pp 194–199
31. Bercnguer C, Grall A, Dieulle L, Roussignol M (2003) Maintenance policy for a continuously monitored deteriorating system. Probab. Eng Inf Sci 17:50–235
32. Buijs FA, Hall JW, van Noortwijk JM, Sayers PB (2005) Time-dependent reliability analysis of flood defences using gamma processes. In: Augusti G, Schuëller GI, Ciampoli M (eds.) Safety and reliability of engineering systems and structures, Millpress, pp 2209–2216

33. Dekker R (1996) Applications of maintenance optimization models: a review and analysis. Reliab Eng Syst Safety 51:40–229
34. Dekker R, Scarf PA (1998) On the impact of optimization models in maintenance decision making: the state of the art. Reliab Eng Syst Safety 60:111–119
35. Dickson DM, Waters HR (1993) Gamma processes and finite time survival probabilities. ASTIN Bull 23:259–272
36. Dieulle L, Berenguer C, Grall A, Roussignol M (2003) Sequential condition-based maintenance scheduling for a deteriorating system. Eur J Oper Res 150:61–451
37. Doksum KA, Normand SL (1995) Gaussian models for degradation processes-Part 1: methods for the analysis of biomarker data. Lifetime Data Anal 1:131–144
38. Defrense F, Greber HU, Shiu ESW (1991) Risk theory with the gamma process. ASTIN Bull. 2:177–192
39. Ellingwood BR, Mori Y (1993) Probabilistic methods for condition assessment and life prediction of concrete structures in nuclear power plants. Nucl Eng Des 14:66–155
40. Estes AC, Frangopol DM (1999) Repair optimization of highway bridges using system reliability approach. J Struct Eng 125:766–775
41. Frangopol DM, Kong JS, Gharaibeh ES (2001) Reliability-based life-cycle management of highway bridges. J Comput Civ Eng 15:27–34
42. Golabi K, Kulkarni RB, Way GB (1982) A statewide pavement management system. Interfaces 12:5–21
43. Golabi K, Shepard R (1997) A system for maintenance optimization and improvement of US bridge networks. Interfaces 27:71–88
44. Grall A, Dieulle L, Berenguer C, Roussignol M (2006) Asymptotic failure rate of a continuously monitored system. Reliab Eng Syst Saf 91:126–130
45. Hawk H, Small EP (1998) The BRIDGIT bridge management system. Struct Eng Int 8:309–314
46. Hurley MA (1992) Modelling bedload transport events using an inhomogeneous gamma process. J Hydrol 138:41–529
47. Ishwaran H, James LF (2004) Computational methods for multiplicative intensity models using weighted gamma processes: proportional hazards, marked point processes, and panel count data. J Am Stat Assoc 99:90–175
48. Jia X, Christer AH (2002) A prototype cost model of functional check decisions in reliability-centred maintenance. J Oper Res Soc 53:1380–1384
49. Morcous G (2006) Performance prediction of bridge deck systems using Markov chains. J Perform Constr Facil 20:146–155
50. Newby MJ, Barker CT (2006) A bivariate process model for maintenance and inspection planning. Int J Press Vessels Pip 83:270–275
51. Nicolai RP, Dekker R, van Noortwijk JM (2007) A comparison of models for measurable deterioration: an application to coatings on steel structures. Reliab Eng Syst Saf 92:1635–1650
52. Pandey MD, van Noortwijk JM (2004) Gamma process model for time-dependent structural reliability analysis. In: Proceedings of the second international conference on bridge maintenance, safety and management
53. Pandey MD, Yuan X, van Noortwijk JM(2005) Gamma process model for reliability analysis and replacement of aging structural components. In: Proceedings of ninth international conference on structural safety and reliability
54. Pandey MD, Yuan X, van Noortwijk JM (2009) The influence of temporal uncertainty of deterioration on lifecycle management of structures. Struct Infrastruct Eng 5:145–156
55. Park KS (1988) Optimal continuous-wear limit replacement under periodic inspections. IEEE Trans Reliab 37:97–102
56. Park KS (1988) Optimal wear-limit replacement with wear-dependent failures. IEEE Trans Reliab 37:293–294
57. Park C, Padgett WJ (2005) Accelerated degradation models for failure based on geometric Brownian motion and gamma processes. Lifetime Data Anal 11:511–527
58. Saassouh B, Dieulle L, Grall A (2007) Online maintenance policy for a deteriorating system with random change of mode. Reliab Eng Syst Saf 82:1677–1685

59. Scherer WT, Glagola DM (1994) Markovian models for bridge maintenance management. J Transp Eng 120:37–51
60. Shaked M, Shanthikumar JG (1988) On the first-passage times of pure jump processes. J Appl Prob 25:501–509
61. Shepard RW, Johnson MB (1999) California bridge health index. In: California department of transportation, international bridge management conference IBMC-005, vol I. Denver, Colorado
62. Singpurwalla N (1997) Gamma processes and their generalizations: an overview. Engineering probabilistic design and maintenance for flood protection. Kluwer Academic Publishers, Dordrecht, pp 67–75
63. Sobczyk K (1987) Stochastic models for fatigue damage of materials. Adv Appl Prob 19:652–673
64. Sobczyk K, Spencer BF Jr (1992) Random fatigue: from data to theory. Academic Press, Boston
65. van Beek A, Gaal GCM, van Noortwijk JM, Bakker JD (2003) Validation model for service life prediction of concrete structures. Second international RILEM workshop on life prediction and aging management of concrete structures, Paris, France
66. van der Weide H (1997a) Gamma processes. In: Cooke M, Medel M, Vrijling H (eds) Engineering probabilistic design and maintenance for flood protection. Kluwer, Dordrect, pp 77–83
67. van Noortwijk JM, Kok M, Cooke RM (1997b) Optimal maintenance decisions for the sea-bed protection of the Eastern-Scheldt barrier. In: Cooke M, Medel M, Vrijling H (eds) Engineering probabilistic design and maintenance for flood protection. Kluwer, Dordrect, pp, 25–56
68. van Noortwijk JM, Klatter HE (1999) Optimal inspection decisions for the block mats of the Eastern-Scheldt barrier. Reliab Eng Syst Saf 65:203–211
69. van Noortwijka JM, Frangopol DM (2004) Two probabilistic life-cycle maintenance models for deteriorating civil infrastructures. Prob Eng Mech 19:345–359
70. van Noortwijk JM, Kallen MJ, Pandey MD (2005) Gamma processes for time-dependent reliability of structures. In: Proceedings of european safety and reliability conference
71. van Noortwijk JM, van der Weide JAM, Kallen MJ, Pandey MD (2007) Gamma processes and peaks-over-threshold distributions for time-dependent reliability. Reliab Eng Syst Saf 92:1651–1658
72. Wang H (2002) A survey of maintenance policies of deteriorating systems. Eur J Oper Res 139(4):69–89
73. Wang W, Scarf PA, Smith MJ (2000) On the application of a model of condition-based maintenance. J Oper Res Soc 51:1218–1227
74. Wenocur MLA (1989) Reliability model based on the gamma process and its analytic theory. Adv Appl Prob 21:899–918
75. Wirahadikusumah R, Abraham D, Iseley T (2001) Challenging issues in modeling deterioration of combined sewers. J Infrastruct Syst 7:77–84
76. Whitmore GA (1995) Estimating degradation by a Wiener diffusion process subject to measurement error. Lifetime Data Anal 1:19–307
77. Yang S, Frangopol DM, Neves LC (2004) Service life prediction of structural systems using lifetime functions with emphasis on bridges. Reliab Eng Syst Saf 86:39–51
78. Yang Y, Klutke G-A (2000) Lifetime-characteristics and inspections schemes for Levy degradation processes. IEEE Trans Reliab 49:82–377

Chapter 3
Storage Models: Control of Dams Using $\mathbf{P}_{\lambda,\tau}^{M}$ Policies

Abstract We discuss the problem of control of a dam using $\mathrm{P}_{\lambda,\tau}^{M}$ control policies when the input process is a subordinator, a spectrally positive Lévy process, and a spectrally positive Lévy process reflected at its infimum. We describe the content process by hitching Lévy processes and spectrally positive Lévy processes reflected at the full capacity of the dam, killed at the times of first up crossing and down crossing of levels λ and τ, respectively. Using the theory and methods of scale functions of spectrally Lévy processes, we give expressions on the first passage problem of the dam content. The potential measures for the content process up to the time of first passage through levels λ and τ are obtained. Using these results we find the total discounted as well as the long-run average costs.

Keywords $P_{\lambda\tau}^{M}$ policies · Spectrally positive Lévy processes · Spectrally positive Lévy processes reflected at its infimum · Scale functions · Exit times · α-potentials · Total discounted and long-run-average costs

3.1 Introduction and Summary

In this chapter, we deal with the control of a finite dam, using $\mathrm{P}_{\lambda,\tau}^{M}$ control policies. In these policies, the water release rate is assumed to be zero, until the water level reaches level $\lambda(\lambda > 0)$ from there on the water is released at a rate $M(M > 0)$, until it reaches level τ, $\tau < \lambda$. Once level τ is reached, the release rate stays at zero till level λ is reached again, and so on. We treat the cases where the input process is a subordinator, a spectrally positive Lévy process, and a spectrally positive Lévy process reflected at its infimum. In [1], the authors consider the control of a finite dam, with capacity $V > 0$, where the water input is a Wiener process, using $P_{\lambda,\tau}^{M}$ policies. They use the total discounted as well as the long-run average cost criteria. In [2], similar results are obtained using simpler methods. In [3], the case where the input process is a compound Poisson process is considered, and the authors obtain the long-run average cost case using the $P_{\lambda,0}^{M}$ policy. In [4], the $P_{\lambda,0}^{M}$ policy used to assess the workload of an M/G/1 queuing system. In [5], the log-run average cost for $P_{\lambda,\tau}^{M}$ policy in a finite dam, when the input process is a compound Poisson process, with a negative drift term, is treated. In [6], the author treats the case where the water input is a compound Poisson process with a positive drift, using the total discounted as well as the long-run average cost criteria. In [7], he extends the results obtained in [6], where he assumes that the input process is subordinator. In [8], he treats the more general cases where the input process is assumed to be a spectrally positive Lévy or a spectrally positive Lévy process reflected at its infimum. In [9], optimal management of a dam where the water input is a continuous-time controlled Markov chain is discussed.

This chapter consists of six sections. In Sect. 3.2, we give basic definitions and results that are needed to compute the total discounted as well as the long-run average costs of running the dam. In Sect. 3.3, the $\mathrm{P}_{\lambda,\tau}^{M}$ control policies and the associated cost functionals are discussed in details. In Sect. 3.4, expressions on the first passage problems (above level λ and below level τ) for the dam content (water level) are obtained, the potentials of the processes obtained by killing the content process at these times are also obtained, and the cost functionals are computed, assuming that the input process is a subordinator. Section 3.5, gives results parallel to the results obtained in Sect. 3.4, when that the input process is a spectrally positive Lévy process and a spectrally positive Lévy process reflected at its infimum. Special cases are discussed and some previous results are recovered in Sect. 3.6.

3.2 Basic Definitions and Results

In this section, we discuss some basic definitions and results dealing with regenerative processes. These results will be used in the rest of this chapter.

Definition 3.1 A stochastic process $X = \{X_t, t \geq 0\}$ is called *regenerative*, if there exists a renewal process $S = \{S_n, n \geq 0, S_0 = 0\}$, with independent identically distributed cycle lengths $\{W_n = S_n - S_{n-1}, n \geq 1\}$ such that for $n \geq 1$, the processes $\{X_{S_{n-1}+t}, 0 \leq t \leq W_n\}$ are independent and have the same probability distribution.

Definition 3.2 A stochastic process $X = \{X_t, t \geq 0\}$ is called *delayed regenerative* if there exists a renewal process $S = \{S_n, n \geq 1\}$ and a positive random variable $S_0 < S_1$, $P\{S_0 > 0\} > 0$, such that the process $\{X_t, 0 \leq t \leq S_0\}$ is independent of the process $\{X_t, t > S_0\}$; furthermore, the latter process is regenerative with regeneration cycles $\{W_n = S_n - S_{n-1}, n \geq 1\}$.

Intuitively, a delayed regenerative process is a process in which the first cycle has a different distribution from (but is still independent of) the independent identically distributed cycles following it.

Theorem 3.3 Let $X = \{X_t, t \geq 0\}$ be a delayed regenerative process with state space $E \subset R$, and regeneration times $S_0, S_1, \ldots$. Define, $W_0 = S_0$, and $W_n = S_n - S_{n-1}$, $n \geq 1$. Assume that $X_{S_n} = y$ almost everywhere, $n = 0, 1, \ldots$ Let $C : E \to R$, C is bounded. For $\alpha \in R_+$, $n = 0, 1, \ldots$, let

$$C_\alpha(n) = \int_0^{W_n} e^{-\alpha t} C(X_t) dt,$$
$$C(n) = \int_0^{W_n} C(X_t) dt.$$

Then, for $x \in E$

(i)

$$E_x \int_0^\infty e^{-\alpha t} C(X_t) dt = E_x[C_\alpha(0)] + \frac{E_x[e^{-\alpha W_0}] E_y[C_\alpha(1)]}{1 - E_y[e^{-\alpha W_1}]}. \tag{3.1}$$

(ii) if $E_x[S_0] < \infty$, and $E_y[W_1] < \infty$

$$\lim_{t\to\infty} \frac{1}{t}[\int_0^t C(X_s) ds] = \frac{E_y[C(1)]}{E_y[W_1]}, \text{ almost surely } P_x. \tag{3.2}$$

Proof (i) The proof of (i) follows in a manner similar to the proof of Theorem 2.29 with obvious modifications, and hence is omitted.

(ii) The proof follows from Theorem 3.6.1 (i) of [10] since, $E_y[\int_0^{W1} C(X_t) dt] \leq \|C\| E_y[W_1] < \infty$, and

$$\lim_{t\to\infty} \frac{1}{t} \int_0^t C(X_t) dt = \lim_{t\to\infty} \frac{1}{t}[\int_0^{S_0} C(X_t) dt + \int_{S_0}^t C(X_t) dt]$$
$$= \lim_{t\to\infty} \frac{1}{t} \int_{S_0}^t C(X_t) dt,$$

and the process $\{X_t, t > S_0\}$, is regenerative with regeneration cycle lengths $\{W_n = S_n - S_{n-1}, n \geq 1\}$. ■

A positive random variable Y is said to be *lattice* if there exists $d \geq 0$ such that Y takes on integral multiples of d.

The following corollary gives the stationary distribution of the process X.

Corollary 3.4 Let X be a delayed regenerative process with state space E satisfying the assumptions of Theorem 3.3. Assume that, for each $x \in E$, $E_x[S_0] < \infty$, W_1 is nonlattice, and $E_y[W_1] < \infty$. Then, for any Borel set $B \subset E$, and each $x \in E$

$$\lim_{t\to\infty} P_x\{X_t \in B\} = \frac{E_y[\int_0^{W1} \mathbf{I}_B(X_t)dt]}{E_y[W_1]}. \tag{3.3}$$

Proof The proof follows from (3.1) by taking $C(x) = \mathbf{I}_B(x)$, using the *Bounded Convergence Theorem*, and since $\lim\limits_{t\to\infty} P_x\{X_t \in B\} = \lim\limits_{\alpha\to 0}\{\alpha E_x \int_0^\infty e^{-\alpha t}\mathbf{I}_B(X_t)dt\}$. ■

3.3 The $\mathrm{P}^{M}_{\lambda,\tau}$ Control Policies and Their Associated Cost Functionals

We consider the control of a finite dam, using $P^{M}_{\lambda,\tau}$ policies. In these policies, over time, the water input (denoted by $I = (I_t, t \geq 0)$ is a stochastic process and the release rate is assumed to be zero until the water reaches level $\lambda > 0$; as soon as this happens, the water is released at rate $M > 0$ until the water content reaches level $\tau > 0$, $\lambda > \tau$. The release rate is increased from 0 to M with a starting cost K_1M, or decreased from M to zero with a closing cost K_2M. Moreover, for each unit of output, a reward R is received. Furthermore, there is a penalty cost which accrues at a rate f, where f is a bounded measurable function. Throughout, we will denote the capacity of the dam by V, $V > 0$. For $t \geq 0$, we define $N_t = I_t - Mt$, and let $N = \{N_t, t \geq 0\}$.

For each $t \in R_+$, let Z_t be the dam content at time t, $Z = \{Z_t, t \in R_+\}$. We define the following sequence of stopping times:

$$\hat{T}_0 = \inf\{t \geq 0 : Z_t \geq \lambda\}, \quad \overset{*}{T}_0 = \inf\{t \geq \hat{T}_0 : Z_t \leq \tau\},$$

$$\hat{T}_n = \inf\{t \geq \overset{*}{T}_{n-1} : Z_t \geq \lambda\}, \quad \overset{*}{T}_n = \inf\{t \geq \hat{T}_n : Z_t \leq \tau\}, \quad n = 1, 2, ...$$

It follows that the process Z is a delayed regenerative process with regeneration times $\{\overset{*}{T}_n, n = 0, 1, ...\}$, and its state space is the interval $(l, V]$. During a given cycle, the release rate is either 0 or M. When the release rate is zero, the process Z has the same distribution as the input process and remains so till the water reaches level λ; from then until it reaches level τ, the content process behaves like the process N reflected at V, we denote this process by $\overset{*}{I}$. It follows that, for each $t \geq 0$,

$$\overset{*}{I}_t = N_t - \sup_{0\leq s\leq t}((N_t - V) \vee 0). \tag{3.4}$$

The penalty cost rate function is defined as follows:

$$f(z,r) = \begin{cases} g(z) & (z,r) \in (l,\lambda) \times \{0\} \\ g^*(z) & (z,r) \in (\tau,\infty) \times \{M\} \end{cases}$$

where $g : (l,\lambda) :\to R_+$, $g^* : (\tau, V] :\to R_+$ are bounded measurable functions.

The types of input processes that we deal with in this chapter enjoy the property that, for $n = 0, 1, ..., Z_{\overset{*}{T}_n} = \tau$ almost everywhere, and the random variables describing the lengths of the successive cycles are nonlattice.

For $x \in (l, V]$, and $\alpha \in R_+$, the expected total discounted and nondiscounted costs during the interval $[0, \hat{T}_0)$, denoted by $C^\alpha_g(x)$ and $C_g(x)$, respectively, are given as follows:

$$C^\alpha_g(x) = E_x[\int_0^{\hat{T}_0} e^{-\alpha t} g(I_t)dt], \tag{3.5}$$

$$C_g(x) = E_x[\int_0^{\hat{T}_0} g(I_t)dt]. \tag{3.6}$$

Let $\boldsymbol{C}^\alpha_{g^*}, \boldsymbol{C}_{g^*}$ denote the expected total discounted and nondiscounted costs during the interval $[\hat{T}_0, \overset{*}{T}_0)$, respectively. It follows that, for $x \in (\lambda, V]$,

$$\boldsymbol{C}^\alpha_{g^*}(x) = E_x[\int_0^{\overset{*}{T}_0} e^{-\alpha t} g^*(\overset{*}{I}_t)dt], \tag{3.7}$$

$$\boldsymbol{C}_{g^*}(x) = E_x[\int_0^{\overset{*}{T}_0} g^*(\overset{*}{I}_t)dt]. \tag{3.8}$$

We now discuss the computations of the cost functionals using the total discounted cost as well as the long-run average cost criteria. Let f be the cost rate function, $\alpha \geq 0$ be the discounting factor. Given $Z_0 = x$, we denote the total discounted and the long-run average costs associated with a $P^M_{\lambda,\tau}$ control policy, by $\mathbb{C}_\alpha(x,\lambda,\tau)$ and $\mathbb{C}(\lambda,\tau)$, respectively. For $x \in (l, V]$, they are defined as follows:

$$\mathbb{C}_\alpha(x,\lambda,\tau) = E_x[\int_0^\infty e^{-\alpha t} f(Z_t)dt, \tag{3.9}$$

$$\mathbb{C}(\lambda,\tau) = \lim_{t\to\infty} \frac{1}{t} \int_0^t e^{-\alpha t} f(Z_s)ds. \tag{3.10}$$

Let $C_\alpha(x)$ and $C(x)$ be the expected discounted and nondiscounted costs during the interval $[0, \overset{*}{T}_0)$, when $Z_0 = x$, respectively. From (3.1) and (3.2), it follows that

$$\mathbb{C}_\alpha(\lambda,\tau) = C_\alpha(x) + \frac{E_x[\exp(-\alpha \overset{*}{T}_0)]C_\alpha(\tau)}{1 - E_\tau[\exp(-\alpha \overset{*}{T}_0)]}, \tag{3.11}$$

$$\mathbb{C}(\lambda,\tau) = \frac{C(\tau)}{E_\tau[\overset{*}{T}_0]}, \quad \text{if} \quad E_\tau[\overset{*}{T}_0] < \infty. \tag{3.12}$$

The following lemma establishes how $C_\alpha(x)$ and $C(x)$ can be computed.

Lemma 3.5 The expected discounted and nondiscounted costs during the interval $[0, \overset{*}{T}_0)$, when $Z_0 = x$ are computed as follows:

(i) For $x \in (\lambda, V]$

$$C_\alpha(x) = M\{K_1 - RE_x \int_0^{\overset{*}{T}_0} e^{-\alpha t} dt\} + \boldsymbol{C}^\alpha_{g^*}(x). \tag{3.13}$$

(ii) For $x \in (l, \lambda]$

$$C_\alpha(x) = M\{K_2 + K_1 E_x[e^{-\alpha \hat{T}_0}] - \frac{R}{\alpha}[E_x[e^{-\alpha \hat{T}_0}] - E_x[e^{-\alpha \overset{*}{T}_0}]]\}$$
$$+ C^\alpha_g(x) + E_x[e^{-\alpha \hat{T}_0} \boldsymbol{C}^\alpha_{g^*}(I_{\hat{T}_0} \wedge V)]. \tag{3.14}$$

(iii) For $x \in (l, \lambda]$

$$C(x) = M\{K + R(E_x[\hat{T}_0] - E_x[\overset{*}{T}_0]) + C_g(x) + E_x[\boldsymbol{C}_{g^*}(I_{\hat{T}_0} \wedge V)]. \tag{3.15}$$

Proof (i) The proof is immediate from the definition of the $P^M_{\lambda,\tau}$ control policy.

(ii) To compute $C_\alpha(x)$ for for $x \in (l, \lambda]$, we let F be the sigma algebra generated by $(\hat{T}_0, Z_{\hat{T}_0})$ and proceed as follows:

$$C_\alpha(x) = M\{K_2 + K_1 E_x[e^{-\alpha \hat{T}_0}] - RE_x \int_{\hat{T}_0}^{\overset{*}{T}_0} e^{-\alpha t} dt\}$$
$$+ E_x[\int_0^{\hat{T}_0} e^{-\alpha t} g(Z_t) dt] + E_x[\int_{\hat{T}_0}^{\overset{*}{T}_0} e^{-\alpha t} g^*(Z_t) dt]$$
$$= M\{K_2 + K_1 E_x[e^{-\alpha \hat{T}_0}] - \frac{R}{\alpha}(E_x[e^{-\alpha \hat{T}_0}] - E_x[e^{-\alpha \overset{*}{T}_0}])\}$$
$$+ E_x[\int_0^{\hat{T}_0} e^{-\alpha t} g(Z_t) dt] + E_x[\int_{\hat{T}_0}^{\overset{*}{T}_0} e^{-\alpha t} g^*(Z_t) dt]$$

$$= M\{K_2 + K_1 E_x[e^{-\alpha\hat{T}_0}] - \frac{R}{\alpha}(E_x[e^{-\alpha\hat{T}_0}] - E_x[e^{-\alpha\overset{*}{T}_0}])\}$$

$$+ C_g^{\alpha}(x) + E_x[E_x[\int_{\hat{T}_0}^{\overset{*}{T}_0} e^{-\alpha t} g^*(Z_t)dt|F]]$$

$$= M\{K_2 + K_1 E_x[e^{-\alpha\hat{T}_0}] - \frac{R}{\alpha}(E_x[e^{-\alpha\hat{T}_0}] - E_x[e^{-\alpha\overset{*}{T}_0}])\}$$

$$+ C_g^{\alpha}(x) + E_x[e^{-\alpha\hat{T}_0} E_{I_{\hat{T}_0}\wedge V}[\int_0^{\overset{*}{T}_0} e^{-\alpha t} g^*(I_t^*)dt]]$$

$$= M\{K_2 + K_1 E_x[e^{-\alpha\hat{T}_0}] - \frac{R}{\alpha}[E_x[e^{-\alpha\hat{T}_0}] - E_x[e^{-\alpha\overset{*}{T}_0}]]\}$$

$$+ C_g^{\alpha}(x) + E_x[e^{-\alpha\hat{T}_0} \boldsymbol{C}_{g^*}^{\alpha}(I_{\hat{T}_0} \wedge V)],$$

where the third equation follows from the definition of $C_g^{\alpha}(.)$, the fourth equation follows from the definition of the content process Z, and since $e^{-\alpha\hat{T}_0} \in F$. The fifth equation follows from the definition of $\boldsymbol{C}_{g^*}^{\alpha}(.)$.

(iii) The proof follows by letting $\alpha \to 0$, in (3.14). ■

The following theorem gives the stationary distribution of the process Z.

Theorem 3.6 Let Z be the content process described above, and suppose that $Z_{\overset{*}{T}_0} = \tau$ almost everywhere. Assume that $\overset{*}{T}_0$ is nonlattice, and for every $x \in E$, $E_x[\overset{*}{T}_0] < \infty$. For $x \in E$, and any Borel set $B \subset E$, define $h(x) = \mathbf{I}_B(x)$, then

$$\lim_{t\to\infty} P_x\{Z_t \in B\} = \frac{C_h(\tau) + E_\tau[\boldsymbol{C}_h(I_{\hat{T}_0} \wedge V)]}{E_\tau[\overset{*}{T}_0]}. \tag{3.16}$$

Proof The proof follows from (3.3), since starting at τ, $\overset{*}{T}_0$ has the same distribution as $W_1 \overset{\text{def}}{=} \overset{*}{T}_1 - \overset{*}{T}_0$, and $Z_{\overset{*}{T}_0} = \tau$ almost surely implies that $Z_{\overset{*}{T}_n} = \tau, n = 0, 1, ..$ almost surely. ■

3.4 Subordinator Input

In this section, we discuss the computations of the entities involved in the total discounted and the long-run average costs given in (3.11) and (3.12), when the input process I is a subordinator with Lévy measure ν, and Laplace exponent ψ, in the sense defined in of Sect. 1.4.

For $\alpha \in R_+$, let $\mathbf{R}^\alpha$ and U^α be the potential of the processes I and the process obtained by killing the process I at time $\hat{T}_0$, respectively. Denote $\mathbf{R}^0$ and U^0 by $\mathbf{R}$ and U, respectively. From (1.40), for $x \in [0, \lambda)$, C_g^α and C_g given in (3.5) and (3.6) can be expressed as follows:

$$C_g^\alpha(x) = U^\alpha g(x) = \int_{[0,\lambda)} g(y)\mathbf{R}^\alpha(x, dy),$$

$$C_g(x) = Ug(x) = \int_{[0,\lambda)} g(y)\mathbf{R}(x, dy).$$

Note that for each $x, y \in R_+$, $\mathbf{R}^\alpha(x, dy) = \mathbf{R}^\alpha(0, dy - x)$, we denote $\mathbf{R}^\alpha(0, dy)$ by $\mathbf{R}^\alpha(dy)$ throughout. We note that, for $y < 0$, $\mathbf{R}^\alpha(dy) = 0$.

The following lemma will be used extensively throughout this chapter.

Lemma 3.7 Let $S = \{S_t, t \geq 0\}$ be a strong Markov process. Define, $\mathcal{G} = \{\sigma(S_u, u \leq t)\}_{t \geq 0}$, κ to be any stopping time with respect to $\mathcal{G}$, and U^α as the α-potential of the process S killed at κ. Denote the state space of this process by E. Then, for $x \in E$

$$E_x[e^{-\alpha\kappa}] = 1 - \alpha U^\alpha \mathbf{I}_E(x). \tag{3.17}$$

Proof From the definition of U^α and for any bounded measurable function f whose domain is E, we have

$$U^\alpha f(x) = E_x[\int_0^\kappa e^{-\alpha t} f(S_t)dt] = \int_E f(y)U^\alpha(x, dy).$$

Taking f to be identically equal to one, we have

$$\frac{1 - E_x[e^{-\alpha\kappa}]}{\alpha} = U^\alpha \mathbf{I}_E(x).$$

The required result is immediate from the last equation above. ∎

The following lemma is immediate from Lemmas 1.22 and 3.7.

Lemma 3.8 For $x \in [0, \lambda)$

$$E_x(\exp(-\alpha \hat{T}_0)) = 1 - \alpha \mathbf{R}^\alpha \mathbf{I}_{[0,\lambda-x)}(0) = \alpha \mathbf{R}^\alpha \mathbf{I}_{[\lambda-x,\infty)}(0), \tag{3.18}$$

$$E_x(\hat{T}_0) = \mathbf{R}\mathbf{I}_{[0,\lambda-x)}(0). \tag{3.19}$$

The following is a restatement of Eq. (8) of [11], the proof is outside the scope of this book and is omitted.

Lemma 3.9 For $x \in R_+$, we let $T_x^+ = \inf\{t \geq 0 : I_t \geq x\}$, then for α, β,

$$E_0[\exp(-\beta T_x^+ - \alpha I_{T_x^+})] = (\beta + \psi(\alpha)) \int_{[x,\infty)} \exp(-\alpha z)\mathbf{R}^\beta(dz). \tag{3.20}$$

The following Lemma gives the Laplace transform of $I_{\hat{T}_0}$ as well as the expected value of $I_{\hat{T}_0}$.

Lemma 3.10 (i) Let ψ be the Laplace exponent of the input process I. For $x \in [0, \lambda)$ and $\alpha \in R_+$, we have

$$E_x[\exp(-\alpha I_{\hat{T}_0})] = \exp(-\alpha x)[1 - \psi(\alpha) \int_{[0,\lambda-x)} \exp(-\alpha z)\mathbf{R}(dz)]. \tag{3.21}$$

(ii) For $x \in [0, \lambda)$

$$E_x(I_{\hat{T}_0}) = x + E_0(I_1)\mathbf{R}\mathbf{I}_{[0,\lambda-x)}(0). \tag{3.22}$$

Proof (i) Let T_x^+ be as defined in Lemma 3.9. From the definition of $\hat{T}_0$, since the process I is a Lévy process, for $x \in [0, \lambda)$ and $\alpha \geq 0$, we have

$$\begin{aligned}
E_x[\exp(-\alpha I_{\hat{T}_0})] &= E_0[\exp(-\alpha(x + I_{T^+_{\lambda-x}}))] \\
&= \exp(-\alpha x)[\psi(\alpha) \int_{[\lambda-x,\infty)} \exp(-\alpha z)\mathbf{R}(dz)] \\
&= \exp(-\alpha x)[\psi(\alpha)\{ \int_{[0,\infty)} \exp(-\alpha z))\mathbf{R}(dz) \\
&\quad - \int_{[0,\lambda-x)} \exp(-\alpha z))\mathbf{R}(dz)\}] \\
&= \exp(-\alpha x)[\psi(\alpha)\{\frac{1}{\phi(\alpha)} - \int_{[0,\lambda-x)} \exp(-\alpha z))\mathbf{R}(dz)\}] \\
&= \exp(-\alpha x)[1 - \psi(\alpha) \int_{[0,\lambda-x)} \exp(-\alpha z))\mathbf{R}(dz)]
\end{aligned}$$

where the second equation follows from Lemma 3.9 by letting $\beta \to 0$, and the fourth equation follows from the definition of $\psi(\alpha)$ and $\mathbf{R}$.

(ii) For $y \in R_+$, we let T_y^+ be as defined in the proof of (i) above. For $x \in [0, \lambda)$, we have

$$\begin{aligned}
E_x(I_{\hat{T}_0}) &= x + E_0(I_{T^+_{\lambda-x}}) \\
&= x + \lim_{\alpha\to 0} [\frac{1 - E_0[\exp(-\alpha I_{T^+_{\lambda-x}})]}{\alpha}]
\end{aligned}$$

$$
\begin{aligned}
&= x + \lim_{\alpha \to 0} \frac{\psi(\alpha)}{\alpha} \int_{[0,\lambda-x)} \exp(-\alpha z))\mathbf{R}(dz)] \\
&= x + \psi^{'}(0)\mathbf{R}\mathbf{I}_{[0,\lambda-x)}(0) \\
&= x + E_0(I_1)\mathbf{R}\mathbf{I}_{[0,\lambda-x)}(0)
\end{aligned}
$$

where the third equation follows from (3.21), the fourth equation follows because $\psi(0) = 0$, and the fifth equation follows since $\psi^{'}(0) = E_0(I_1)$. ■

The following theorem extends Theorem 1.23, and it is needed to compute the last entity in (3.14).

Theorem 3.11 Let I be the subordinator input process with a Lévy measure υ. For $\alpha \geq 0$, let $\mathbf{R}^\alpha$ be its α-potential. For each, $x \leq \lambda$, $z \geq \lambda$,

$$
E_x[e^{-\alpha \hat{T}_0}, I_{\hat{T}_0} \in dz] = \int_{(0,\lambda]} \upsilon(dz - y)\mathbf{R}^\alpha(dy - x). \tag{3.23}
$$

Proof For $t \in R_+$, we define, $\Delta I_t = I_t - I_{t-}$. Let M be the Poisson random measure associated with the jumps of I. For $x < \lambda$, $\alpha \geq 0$, $C \subset [\lambda, \infty)$, and $D \subset (0, \lambda)$, we have

$$
\begin{aligned}
&E_x[e^{-\alpha \hat{T}_0}, I_{\hat{T}_0} \in C, I_{\hat{T}_0 -} \in D] \\
&\quad = E_x[e^{-\alpha \hat{T}_0}, \Delta I_{\hat{T}_0} \in C - I_{\hat{T}_0 -}, I_{\hat{T}_0 -} \in D] \\
&\quad = E_x[\int_{[0,\infty)\times(0,\infty)} e^{-\alpha t}\mathbf{I}_{\{I_t - \in D\}}\mathbf{I}_{\{y \in C - I_t -\}} M(dt, dy)] \\
&\quad = E_x[\int_{[0,\infty)} e^{-\alpha t}\mathbf{I}_{\{I_t \in D\}}\nu(C - I_t)dt] \\
&\quad = E_x[\int_{[0,\infty)\times D} e^{-\alpha t}\nu(C - y)\mathbf{I}_{\{I_t \in dy\}} dt dy] \\
&\quad = \int_D \nu(C - y)\mathbf{R}^\alpha(x, dy),
\end{aligned}
$$

where the fourth equation follows from the *compensation formula* (Theorem 1.16). Our assertion is proved by taking $D = [0, \lambda]$, and since $\mathbf{R}^\alpha(x, dy) = \mathbf{R}^\alpha(dy - x)$. ■

Now we turn our attention to computing $C^\alpha_{g^*}(x)$, $E_x[\exp(\overset{*}{T}_0)]$, and $E_x[\overset{*}{T}_0]$, when $x \in [\lambda, V]$. Let $\eta_\tau = \inf\{t \geq 0 : \overset{*}{I}_t \leq \tau\}$. Denote the process obtained by killing the process $\overset{*}{I}$, at time η_τ by $\overset{*}{X}$, i.e., for each $t \geq 0$,

$$\overset{*}{X}_t = \{\overset{*}{I}_t, t < \eta_\tau\},$$

note that the process $\overset{*}{X}$ has the interval $(\tau, V]$ as its state space, and let $\overset{*}{U}^{\alpha}$ be its α-potential. Let $\boldsymbol{C}^{\alpha}{}_{g^*}$ and $\boldsymbol{C}_{g^*}$ be as defined in (3.7) and (3.8), respectively. Then, for each $x \in (\tau, V]$

$$\boldsymbol{C}^{\alpha}{}_{g^*}(x) = \overset{*}{U}^{\alpha} g^*(x), \tag{3.24}$$

$$\boldsymbol{C}_{g^*}(x) = \overset{*}{U} g^*(x). \tag{3.25}$$

The following is a well-known result (see Theorem 8.11 of [12]), its proof is rather difficult, and is outside the scope of this book.

Lemma 3.12 Let X be a spectrally negative Lévy process reflected at its infimum, and assume that Y is a spectrally negative Lévy process reflected at its supremum. Let $W^{(\alpha)}$ and $Z^{(\alpha)}$, be the corresponding α-scale and adjoint α-scale functions, respectively. For $a \geq 0$, let U^{α}, $\overset{-}{U}^{\alpha}$ be the α-potential of the processes obtained by killing X and Y at the time of first crossing level a from below, respectively. Then,

(i) U^{α} is absolutely continuous with respect to the Lebesgue measure on $[0, a]$, its density is given by

$$u^{\alpha}(x, y) = \frac{Z^{(\alpha)}(x) W^{(\alpha)}(a - y)}{Z^{(\alpha)}(a)} - W^{(\alpha)}(x - y), \; x, y \in [0, a]. \tag{3.26}$$

(ii) For any $x, y \in [0, a)$,

$$\overset{(2)}{U^{\alpha}}(x, dy) = \frac{W^{(\alpha)}(a - x) W^{(\alpha)}(dy)}{W_+^{(\alpha)'}(a)} - W^{(\alpha)}(y - x) dy, \tag{3.27}$$

where for $x, y \in [0, a)$, $W^{(\alpha)}(dy) = W^{(\alpha)}(0)\delta_0(dy) + W_+^{(\alpha)'}(y)dy$, and δ_0 is the delta measure in zero.

Note that the process N, defined in Sect. 3.3, is a spectrally positive Lévy process of bounded variation with Laplace exponent $\phi(\theta) = \theta M - \psi(\theta)$, $\theta \geq 0$. We denote its α-scale and adjoint α-scale functions by $W_M^{(\alpha)}$ and $Z_M^{(\alpha)}$, respectively.

Theorem 3.13 For $\alpha \geq 0$, $\overset{*}{U}^{\alpha}$ is absolutely continuous with respect to the Lebesgue measure on $(\tau, V]$, and a version of its density is given by

$$\overset{*}{u}^{\alpha}(x, y) = \frac{Z_M^{(\alpha)}(V - x) W_M^{(\alpha)}(y - \tau)}{Z_M^{(\alpha)}(V - \tau)} - W_M^{(\alpha)}(y - x), \quad x, y \in (\tau, V]. \tag{3.28}$$

Proof For each $t \geq 0$, we define $B_t = N_t - V$. For any $b \in R$, we define $\gamma_b^+ = \inf\{t \geq 0 : \hat{B}_t - \underline{\hat{B_t}} > b\}$ and $\sigma_b^- = \inf\{t \geq 0 : B_t - \bar{B}_t < b\}$. For Borel set $A \subseteq (\tau, V]$ and $x \in (\tau, V]$, we have

$$\begin{aligned} P_x\{\overset{*}{X}_t \in A\} &= P_x\{\overset{*}{I}_t \in A, t < \eta_\tau\} \\ &= P_x\{N_t - \sup_{s \leq t}((N_s - V) \vee 0) \in A, t < \eta_\tau\} \\ &= P_{x-V}\{B_t - \bar{B}_t \in A - V, t < \sigma^-_{\tau - V}\} \\ &= P_{V-x}\{\hat{B}_t - \underline{\overset{\wedge}{B_t}} \in V - A, t < \gamma^+_{V-\tau}\} \end{aligned}$$

Our assertion follows immediately from Lemma 3.12 (i). ■

The following theorem gives Laplace transform of the distribution of the stopping time $\overset{*}{T}_0$ and $E_x[\overset{*}{T}_0]$, $x \in (\tau, V]$.

Theorem 3.14 (i) Let $x \in (\tau, V]$ and $\alpha \in R_+$, then we have

$$E_x[e^{-\alpha\eta_\tau}] = \frac{Z_M^{(\alpha)}(V - x)}{Z_M^{(\alpha)}(V - \tau)}. \tag{3.29}$$

(ii) For $x \in (\tau, V]$

$$E_x[\eta_\tau] = \bar{W}_M(V - \tau) - \bar{W}_M(V - x), \tag{3.30}$$

where, $\bar{W}_M(x) = \int_0^x W_M(y)dy.$

Proof We only prove (i), the proof of (ii) follows easily from (i) and is omitted. For $x \in (\tau, V]$, we have

$$\begin{aligned} E_x[e^{-\alpha\eta_\tau}] &= 1 - \alpha \overset{*\alpha}{U} \mathbf{I}_{(\tau,V]}(x) \\ &= 1 - \alpha \int_{(\tau,V]} \overset{*\alpha}{U}(x, dy) \\ &= 1 - \alpha \int_\tau^V [\frac{Z_M^{(\alpha)}(V - x) W_M^{(\alpha)}(y - \tau)}{Z_M^{(\alpha)}(V - \tau)} - W_M^{(\alpha)}(y - x))]dy \\ &= 1 - \alpha[\frac{Z_M^{(\alpha)}(V - x)}{Z_M^{(\alpha)}(V - \tau)}\{\frac{Z_M^{(\alpha)}(V - \tau) - 1}{\alpha}\} - \{\frac{Z_M^{(\alpha)}(V - x) - 1}{\alpha}\}] \end{aligned}$$

$$= \frac{Z_M^{(\alpha)}(V-x)}{Z_M^{(\alpha)}(V-\tau)} - Z_M^{(\alpha)}(V-x) + Z_M^{(\alpha)}(V-x)$$
$$= \frac{Z_M^{(\alpha)}(V-x)}{Z_M^{(\alpha)}(V-\tau)},$$

where the first equation follows from (3.17), the third equation follows from (3.28), the fourth equation follows from the definition of the function $Z_M^{(\alpha)}$, and the fifth equation follows the fourth equation after obvious manipulations. ∎

Remark 1 When $V = \infty$, for $\alpha \geq 0$ we let $\eta(\alpha) = \sup\{\theta : \phi(\theta) = \alpha\}$, since $Z^{(\alpha)}(z) = O(e^{\eta(\alpha)z})$ as $z \to \infty$, letting $V \to \infty$, in (3.29) and (3.30), then $x \geq \tau$, we have

$$E_x[e^{-\alpha\eta_\tau}] = e^{-(x-\tau)\eta(\alpha)}, \tag{3.31}$$

$$\begin{aligned} E_x[\eta_\tau] &= (x-\tau)\eta(0)' \\ &= \begin{cases} \frac{x-\tau}{M-E(I_1)} & \text{if } M > E(I_1), \\ \infty & \text{if } M \leq E(I_1). \end{cases} \end{aligned} \tag{3.32}$$

This is consistent with the well-known fact about the busy period of the M/G/1 queuing system.

To compute $E_x[e^{-\alpha \overset{*}{T}_0}]$, $x \in [\tau, V]$, we first note that starting at $x \in [\lambda, V]$, $\overset{*}{T}_0 = \eta_\tau$ almost surely P_x, and hence $E_x[e^{-\alpha \overset{*}{T}_0}]$ is computed using (3.29). We now turn our attention to computing $E_x[e^{-\alpha \overset{*}{T}_0}]$, when $x < \lambda$.

For $x < z,\ z > \lambda$, we denote the right-hand side of (3.23) by $h_\alpha(x, dz)$, i.e.,

$$h_\alpha(x, dz) = \int_{(0,\lambda]} \upsilon(dz - y)\mathbf{R}^\alpha(dy - x).$$

We will denote $h_0(x, dz)$ by $h(x, dz)$, throughout.

Theorem 3.15 Assume that the input process is a subordinator.

(i) for $\alpha \geq 0$, $x < \lambda$

$$E_x[e^{-\alpha \overset{*}{T}_0}] = \frac{1}{Z_M^{(\alpha)}(V-\tau)}\left[\int_{(\lambda,V]} Z_M^{(\alpha)}(V-z)h_\alpha(x, dz) + \int_{(V,\infty)} h_\alpha(x, dz)\right]. \tag{3.33}$$

(ii) for $x < \lambda$

$$E_x[\overset{*}{T}_0] = \mathbf{RI}_{[0,\lambda-x)}(0) + \bar{W}_M(V-\tau) - \int_{(\lambda,V]} \bar{W}_M(V-z)h(x, dz). \tag{3.34}$$

Proof (i) We write

$$
\begin{aligned}
E_x[e^{-\alpha \overset{*}{T}_0}] &= E_x[e^{-\alpha \hat{T}_0 - \alpha(\overset{*}{T}_0 - \hat{T}_0)}] \\
&= E_x[E_x[e^{-\alpha \hat{T}_0 - \alpha(\overset{*}{T}_0 - \hat{T}_0)} \mid \sigma(\hat{T}_0, Z_{\hat{T}_0})] \\
&= E_x[e^{-\alpha \hat{T}_0} \; E_{Z_{\hat{T}_0}} [e^{-\alpha(\overset{*}{T}_0 - \hat{T}_0)}]] \\
&= E_x[e^{-\alpha \hat{T}_0} \; E_{Z_{\hat{T}_0}} [e^{-\alpha \overset{*}{T}_0}]] \\
&= \frac{1}{Z_M^{(\alpha)}(V-\tau)} E_x[e^{-\alpha \hat{T}_0} \; Z_M^{(\alpha)}(V - (I_{T_\lambda^+} \wedge V))] \\
&= \frac{1}{Z_M^{(\alpha)}(V-\tau)} [\int_{(\lambda, V]} Z_M^{(\alpha)}(V - z) h_\alpha(x, dz) + \int_{(V,\infty)} h_\alpha(x, dz)],
\end{aligned}
$$

where the fourth equation follows since, starting at $y \geq \lambda$, $\hat{T}_0 = 0$ almost surely. The fifth equations from (3.29) since $Z_{\hat{T}_0} = I_{\hat{T}_0} \wedge V$ almost surely P_x. The last equation follows from (3.23), since $Z^\alpha(0) = 1$.

(ii) For $x < \lambda$

$$
\begin{aligned}
E_x[\overset{*}{T}_0] &= E_x[\hat{T}_0] + E_x[E_{Z_{\hat{T}_0}} [\overset{*}{T}_0]] \\
&= E_x[\hat{T}_0] + E_x[\bar{W}_M(V-\tau) \; - \; \bar{W}_M(V - (I_{\hat{T}_0} \wedge V))] \\
&= E_x[\hat{T}_0] + \bar{W}_M(V-\tau) \; - \; E_x[\bar{W}_M(V - (I_{\hat{T}_0} \wedge V))],
\end{aligned}
$$

where the second equation follows from (3.30). The assertion is proved using (3.19), (3.23), and the definition of $h(x, dz)$. ■

The following theorem gives similar results when the dam has infinite capacity.

Theorem 3.16 Assume that the input process is a subordinator, with α-potential $\mathbf{R}^\alpha$, and $V = \infty$. Denote $\mathbf{R}^0$ by $\mathbf{R}$, then

(i) for $\alpha \geq 0$, $x < \lambda$

$$
E_x[\exp(-\alpha \overset{*}{T}_0)] = M\eta(\alpha) \exp(-\eta(\alpha)(x-\tau)) \int_{[\lambda - x, \infty)} \exp(-z\eta(\alpha)) \mathbf{R}^\alpha(dz). \tag{3.35}
$$

(ii) for $x < \lambda$

$$E_x[\overset{*}{T}_0] = \begin{cases} (x-\tau) + \frac{M\mathbf{RI}_{[0,\lambda-x)}(0)}{M-E[I_1]}, & \text{if } M > E(I_1) \\ \infty & \text{if } M \leq E(I_1) \end{cases} \tag{3.36}$$

Proof (i) For $y \geq 0$, let T_y^+ be as defined in Lemma 3.9, following similar steps to the ones used in proving (3.33), we have

$$\begin{aligned} E_x[\exp(-\alpha\overset{*}{T}_0)] &= E_x[\exp(-\alpha\hat{T}_0)E_{I_{\hat{T}_0}}\exp(-\alpha\overset{*}{T}_0)] \\ &= E_x[\exp(-\alpha T_\lambda^+)E_{I_{T_\lambda^+}}\exp(-\alpha T_\tau^-)] \\ &= E_x[\exp(-\alpha T_\lambda^+)\exp(-\eta(\alpha)(I_{T_\lambda^+}-\tau))] \\ &= E_0[\exp(-\alpha T_{\lambda-x}^+)\exp(-\eta(\alpha)(I_{T_{\lambda-x}^+}+x-\tau))] \\ &= \exp(-\eta(\alpha)(x-\tau))E_0[\exp(-\alpha T_{\lambda-x}^+ - \eta(\alpha)I_{T_{\lambda-x}^+})] \\ &= (\alpha+\psi(\eta(\alpha))\exp(-\eta(\alpha)(x-\tau))) \int_{[\lambda-x,\infty)} \exp(-z\eta(\alpha))\mathbf{R}^\alpha(dz) \\ &= M\eta(\alpha)\exp(-\eta(\alpha)(x-\tau)) \int_{[\lambda-x,\infty)} \exp(-z\eta(\alpha))\mathbf{R}^\alpha(dz), \end{aligned}$$

where the third equation follows from (3.31), the sixth equation follows from Lemma 3.9, and the last equation follows since $\alpha + \psi(\eta(\alpha)) = M\eta(\alpha)$, as evident from the definition of $\eta(\alpha)$. ■

(ii) For $x < \lambda$, we have for $M > E(I_1)$

$$\begin{aligned} E_x[\overset{*}{T}_0] &= E_x[T_\lambda^+] + E_x[E_{I_{T_\lambda^+}}[T_\tau^-]] \\ &= E[T_{\lambda-x}^+] + E_x[\frac{I_{T_\lambda^+}-\tau}{M-E(I_1)}] \\ &= (x-\tau) + E[T_{\lambda-x}^+] + E[\frac{I_{T_{\lambda-x}^+}}{M-E(I_1)}] \\ &= (x-\tau) + E[T_{\lambda-x}^+] + E[\frac{E[I_1]E[T_{\lambda-x}^+]}{M-E(I_1)}] \\ &= (x-\tau) + \frac{ME[T_{\lambda-x}^+]}{M-E[I_1]} \\ &= (x-\tau) + \frac{M\mathbf{RI}_{[0,\lambda-x)}(0)}{M-E[I_1]} \end{aligned}$$

where the second equality follows from (3.32), the fourth equation follows from (3.22), and the last equation follows from (3.19). The fact that $E_x[\overset{*}{T}_0] = \infty$ if $M \leq E(I_1)$ follows from (3.32) in an obvious manner. ■

Corollary 3.17 Assume that the input process is a subordinator, with α-potential $\mathbf{R}^\alpha$, and $V = \infty$. Assume that $M > E(I_1)$, and let $M^* = M - E(I_1)$. Then, the long-run average cost per a unit of time is as follows:

$$\mathbb{C}(\lambda, \tau) = \frac{KM^* + (M^*/M)[C_g(\tau) + E_\tau[\mathbf{C}_{g^*}(I_{\hat{T}_0})]}{\mathbf{RI}_{[0,\lambda-\tau)}(0)} - RE[I_1]. \tag{3.37}$$

Proof We have

$$\begin{aligned}
\mathbb{C}(\lambda, \tau) &= \frac{M\{K + R(E_\tau[\hat{T}_0] - E_\tau[\overset{*}{T}_0]) + C_g(\tau) + E_\tau[\mathbf{C}_{g^*}(I_{\hat{T}_0})]}{E_\tau[\overset{*}{T}_0]} \\
&= \frac{M\{K - R(\frac{E[I_1]}{M^*} \times E[T^+_{\lambda-\tau}])\} + C_g(\tau) + E_\tau[\mathbf{C}_{g^*}(I_{\hat{T}_0})]}{(ME[T^+_{\lambda-\tau}]/M^*)} \\
&= \frac{KM^* + (M^*/M)[C_g(\tau) + E_\tau[\mathbf{C}_{g^*}(I_{\hat{T}_0})]}{\mathbf{RI}_{[0,\lambda-\tau)}(0)} - RE[I_1],
\end{aligned}$$

where the first equation follows from (3.12) and (3.15), the second equation follows from (3.19), (3.36), and the definition of M^*, while the third equation follows from the second equation and (3.19). ■

Remark 2 Assume that $g(x) = c_1$, $g^*(x) = c_2$, where c_1 and c_2 are positive constants, and the assumptions of Corollary 3.17 are satisfied. From (3.32) and (3.37), it follows that

$$\mathbb{C}(\lambda, \tau) = \frac{M^* K}{\mathbf{RI}_{[0,\lambda-\tau)}(0)} + \frac{1}{M}(c_1 M^* + c_2 E[I_1]) - RE[I_1]. \tag{3.38}$$

3.5 Spectrally Positive and Spectrally Positive Reflected at its Infimum Inputs

In this section, we deal with the case where the input process is either a spectrally positive Lévy process or a spectrally positive Lévy process reflected at its infimum. The computations of the entities involved in the total discounted and the long-run average costs given in (3.11) and (3.12), are discussed.

First we deal with the case where the input process, I, is a spectrally positive Lévy process with Levy measure ν and Laplace exponent ϕ. We will first discuss computations of $C^\alpha_g(x)$, $C_g(x)$, $E_x[e^{-\alpha \hat{T}_0}]$, and $E_x[\hat{T}_0]$. Throughout the rest of this chapter, we let $\hat{I} = -I$, for any $a \in R$, we define $T^-_a = \inf\{t \geq 0 : I_t \leq a\}$,

$\mathsf{T}_a^+ = \inf\{t \ge 0 : \hat{I}_t \ge a\}$, and $\mathsf{T}_a^- = \inf\{t \ge 0 : \hat{I}_t \le a\}$. We will also denote the α-scale and the adjoint α-scale functions of the process I by $W^{(\alpha)}$ and $Z^{(\alpha)}$, respectively.

With the help of (1.43), we are now in a position to find $E_x[e^{-\alpha \hat{T}_0}]$ and $E_x[\hat{T}_0]$.

Theorem 3.18 (i) For $\alpha > 0$ and $x < \lambda$, we have

$$E_x[e^{-\alpha \hat{T}_0}] = Z^{(\alpha)}(\lambda - x) - \frac{\alpha}{\eta(\alpha)} W^{(\alpha)}(\lambda - x). \tag{3.39}$$

(ii) For $x < \lambda$ we have

$$\begin{aligned} E_x[\hat{T}_0] &= \frac{W(\lambda - x)}{\eta(0)} - \bar{W}(\lambda - x), \eta(0) > 0 \\ &= \infty, \quad \eta(0) = 0, \end{aligned} \tag{3.40}$$

where for every $x \ge 0$,

$$\bar{W}(x) = \int_0^x W(y) dy. \tag{3.41}$$

Proof We only prove (i), the proof of (ii) is easily obtained from (i) and hence is omitted. Let U^α be the α-potential of the process X, obtained by killing the process $\hat{I}$ at time $\hat{T}_0$, then

$$\begin{aligned} E_x[e^{-\alpha \hat{T}_0}] &= 1 - \alpha U^\alpha \mathbf{I}_{(-\infty,\lambda)}(x) \\ &= 1 - \alpha \int_{-\infty}^{\lambda} \{W^\alpha(\lambda - x) e^{-(\lambda - y)\eta(\alpha)} - W^\alpha(y - x)\} dy \\ &= 1 + \alpha \int_x^{\lambda} W^\alpha(y - x) dy - \alpha W^\alpha(\lambda - x) \int_{-\infty}^{\lambda} e^{-(\lambda - y)\eta(\alpha)} dy \\ &= Z^{(\alpha)}(\lambda - x) - \frac{\alpha}{\eta(\alpha)} W^{(\alpha)}(\lambda - x), \end{aligned}$$

where the first equation follows from Lemma 3.7, the second equation follows from (3.43), the third equation follows since $W^{(\alpha)}(x) = 0$, $x < 0$, and the last equation follows from the definition of $Z^{(\alpha)}$. ■

We need the following to compute the last term in (3.14).

Proposition 3.19 For $\alpha \ge 0$, let $\overset{(1)}{u^\alpha}(x, y)$ be as given in (1.42), and $x \le \lambda \le z$, then

$$E_x[e^{-\alpha \hat{T}_0}, I_{\hat{T}_0} \in dz, \hat{T}_0 < T_a^-] = \int_a^{\lambda} \upsilon(dz - y) \overset{(1)}{u^\alpha}(x, y) dy, \tag{3.42}$$

Proof Let M be the Poisson random measure associated with the jumps of the process I, and $T = \hat{T}_0 \wedge T_a^-$. For $x < \lambda, \alpha \geq 0, C \subset [\lambda, \infty)$, and $D \subset (a, \lambda)$ we have,

$$E_x[e^{-\alpha \hat{T}_0}, I_{\hat{T}_0} \in C, I_{\hat{T}_0-} \in D, \hat{T}_0 < T_a^-]$$
$$= E_x[\int_{[0,\infty)\times(0,\infty),} e^{-\alpha t} \mathbf{I}_{\{I_{t-}<\lambda, I_{t-}, > a, I_t - \in D\}} \mathbf{I}_{\{y \in C - I_t -\}} M(dt, dy)]$$
$$= E_x[\int_{[0,\infty)} e^{-\alpha t} \mathbf{I}_{\{I_{t-}<\lambda, I_{t-}, > a\}} \mathbf{I}_{\{I_t \in D\}} \nu(C - I_t) dt]$$
$$= E_x[\int_{[0,\infty)} e^{-\alpha t} \mathbf{I}_{\{t<T\}} \mathbf{I}_{\{I_t \in D\}} \nu(C - I_t) dt]$$
$$= E_x[\int_{[0,\infty)} e^{-\alpha t} \mathbf{I}_{\{t<T\}} \nu(C - I_t) \mathbf{I}_{\{X_t \in D\}} dt]$$
$$= E_x[\int_{[0,\infty)\times D} e^{-\alpha t} \mathbf{I}_{\{t<T\}} \nu(C - y) \mathbf{I}_{\{I_t \in dy\}} dt dy]$$
$$= \int_D \nu(C - y) \overset{(1)}{u^\alpha}(x, y) dy,$$

where the third equation follows from the *compensation formula* (Theorem 1.16), and the last equation follows from the definition of $\overset{(1)}{u^\alpha}$. Our assertion is proved by taking $D = (a, \lambda)$. ■

The following corollary gives the formula needed to compute the last term of (3.14) when the input process is a spectrally positive Lévy process.

Corollary 3.20 Let u^α be as defined in (1.43), then, for $\alpha \geq 0$ and for $x \leq \lambda \leq z$,

$$E_x[e^{-\alpha \hat{T}_0}, I_{\hat{T}_0} \in dz\,] = \int_{-\infty}^{\lambda} \upsilon(dz - y) u^\alpha(x, y) dy. \tag{3.43}$$

Proof The proof follows immediately from (1.43) and (3.42) by letting $a \to -\infty$. ■

We note that,

$$C_g^\alpha(x) = \int_{-\infty}^{\lambda} g(y) u^\alpha(x, y) dy,$$
$$C_g(x) = \int_{-\infty}^{\lambda} g(y) u(x, y) dy.$$

We now turn our attention to the case where the input process is spectrally positive reflected at its infimum. In this case, the process killed at time $\hat{T}_0$ has state space $[0, \lambda)$, denote its α-potential by $\overset{(2)}{U}{}^{\alpha}$.

Proposition 3.21 For any $x, y \in [0, \lambda)$,

$$\overset{(2)}{U}{}^{\alpha}(x, dy) = \frac{W^{(\alpha)}(\lambda - x) W^{(\alpha)}(dy)}{W_+^{(\alpha)'}(\lambda)} - W^{(\alpha)}(y - x) dy, \tag{3.44}$$

where $W^{(\alpha)}(dy)$ is defined in Lemma 3.12 (ii).

Proof Note that for each $t \geq 0$,

$$\begin{aligned} I_t &= Y_t - \underline{Y_t} \\ &= \overline{\hat{Y}}_t - \hat{Y}_t, \end{aligned} \tag{3.45}$$

where the process $Y = \{Y_t, t \geq 0\}$ is a spectrally positive Lévy process, with α-scale function $W^{(\alpha)}$. The result follows from Lemma 3.12 (ii), since the process $\hat{Y}$ is a spectrally negative Lévy process. ■

The following results are parallel to (3.39) and (3.40), respectively.

Proposition 3.22 (i) For $\alpha \geq 0$ and $x < \lambda$, we have

$$E_x[e^{-\alpha \hat{T}_0}] = Z^{(\alpha)}(\lambda - x) - W^{(\alpha)}(\lambda - x) \frac{\alpha W^{(\alpha)}(\lambda)}{W_+^{(\alpha)'}(\lambda)}. \tag{3.46}$$

(ii) For $x < \lambda$ we have

$$E_x[\hat{T}_0] = W(\lambda - x) \frac{W(\lambda)}{W_+'(\lambda)} - \bar{W}(\lambda - x). \tag{3.47}$$

Proof The proof of part (i) follows from (3.44) and Lemma 3.7, in a manner similar to the proof of (3.39). The proof of part (ii) follows from part (i) by direct differentiation. ■

Observe that,

$$C_g^{\alpha}(x) = \int_{[0,\lambda)} g(y) \overset{(2)}{U}{}^{\alpha}(x, dy),$$

$$C_g(x) = \int_{[0,\lambda)} g(y) \overset{(2)}{U}(x, dy).$$

To find a formula analogous to (3.21), for the spectrally positive Lévy process reflected at its infimum, we need few definitions.

Define

$$l_\alpha(dz) = W^{(\alpha)}(\lambda - x)\int_{[0,\lambda)} W_+^{(\alpha)'}(dy)\upsilon(dz - y) \qquad (3.48)$$
$$-W_+^{(\alpha)'}(\lambda)\int_0^\lambda dy W^{(\alpha)}(y - x)\upsilon(dz - y), z > \lambda.$$

$$L_\alpha(z) = \int_{(z,\infty)} l_\alpha(du). \qquad (3.49)$$

$$V_\alpha(\lambda) = W_+^{(\alpha)'}(\lambda)Z^{(\alpha)}(\lambda - x) - \alpha W^{(\alpha)}(\lambda - x)W^{(\alpha)}(\lambda). \qquad (3.50)$$

The following proposition gives the required formula.

Proposition 3.23 (i) For $\alpha \geq 0$ and for $x \leq \lambda < z$,

$$E_x[e^{-\alpha\hat{T}_0}, I_{\hat{T}_0} \in dz] = \frac{l_\alpha(dz)}{W_+^{(\alpha)'}(\lambda)}. \qquad (3.51)$$

(ii) For $\alpha \geq 0$

$$E_x[e^{-\alpha\hat{T}_0}, I_{\hat{T}_0} = \lambda] = \frac{V_\alpha(\lambda) - L_\alpha(\lambda)}{W_+^{(\alpha)'}(\lambda)}. \qquad (3.52)$$

Proof (i) Consider the spectrally positive Lévy process $Y = \{Y_t, t \geq 0\}$, given in the proof of Proposition 3.21. For any $a \in R$, we define F_a as the sigma algebra generated by $(Y_s, s \leq t)$, $\tau_a^+ = \inf\{t \geq 0 : Y_t \geq a\}$, $\tau_a^- = \inf\{t \geq 0 : Y_t \leq a\}$, $\sigma_a^+ = \inf\{t \geq 0 : \hat{Y}_t \geq a\}$, and $\sigma_a^- = \inf\{t \geq 0 : \hat{Y}_t \leq a\}$. From (3.45), for $x \geq 0$, $I_0 = x$ if and only if $Y_0 = x$ if and only if $\hat{Y}_0 = -x$. Furthermore, $\hat{T}_0 = \tau_\lambda^+$ and $I_{\hat{T}_0} = Y_{\tau_\lambda^+}$ almost surely on $\{\tau_\lambda^+ < \tau_0^-\}$. Therefore,

$$\begin{aligned} E_x[e^{-\alpha\hat{T}_0}, I_{\hat{T}_0} \in dz] &= E_x[e^{-\alpha\hat{T}_0}, I_{\hat{T}_0} \in dz, \tau_\lambda^+ < \tau_0^-] \\ &\quad + E_x[e^{-\alpha\hat{T}_0}, I_{\hat{T}_0} \in dz, \tau_\lambda^+ \geq \tau_0^-] \\ &= E_x[e^{-\alpha\tau_\lambda^+}, Y_{\tau_\lambda^+} \in dz, \tau_\lambda^+ < \tau_0^-] \\ &\quad + E_x[e^{-\alpha\tau_0^-}, \tau_\lambda^+ \geq \tau_0^-] \times E_0[e^{-\alpha\hat{T}_0}, I_{\hat{T}_0} \in dz] \\ &= E_x[e^{-\alpha\tau_\lambda^+}, Y_{\tau_\lambda^+} \in dz, \tau_\lambda^+ < \tau_0^-] \end{aligned}$$

$$+ E_{-x}[e^{-\alpha\sigma_0^+}, \sigma_{-\lambda}^- \geq \sigma_0^+] \times E_0[e^{-\alpha\hat{T}_0}, I_{\hat{T}_0} \in dz]$$

$$= E_x[e^{-\alpha\tau_\lambda^+}, Y_{\tau_\lambda^+} \in dz, \tau_\lambda^+ < \tau_0^-]$$

$$+ E_{\lambda-x}[e^{-\alpha\sigma_\lambda^+}, \sigma_0^- > \sigma_\lambda^+] \times E_0[e^{-\alpha\hat{T}_0}, I_{\hat{T}_0} \in dz],$$

where the second equation follows from the first equation by conditioning on $F_{\tau_0^-}$ and then using the strong Markov property. The third and fourth equations follow from the definitions of $\hat{Y}, \tau_a^+, \tau_a^-, \sigma_a^+, \sigma_a^-$.

Letting $a \to 0$ in (1.42) and (3.42), we find that the first term in the last equation above is equal to $\int_0^\lambda \nu(dz - y)[W^{(\alpha)}(\lambda - x)\frac{W^{(\alpha)}(y)}{W^{(\alpha)}(\lambda)} - W^{(\alpha)}(y - x)]dy$. From (1.45) the second term is equal to $\frac{W^{(\alpha)}(\lambda-x)}{W^{(\alpha)}(\lambda)}$, and the third term is equal to $\frac{W^{(\alpha)}(\lambda)}{W_+^{(\alpha)'}(\lambda)}\int_0^\lambda dy W_+^{(\alpha)'}(y)\nu(dz - y) - \int_0^\lambda dy W^{(\alpha)}(y)\nu(dz - y)$ (this follows from Theorem 4.1 of [13] by letting the $\beta, \gamma \to 0$).

Our assertion is satisfied by replacing each of the three terms in the last equation by the corresponding value indicated above and after some algebraic manipulations, which we omit.

(ii) The proof is immediate from (3.46) and (3.51). ■

Now we turn our attention to computing $C_\alpha(x, \tau, M)$, $E_x[\exp(-\alpha\overset{*}{T}_0)]$, and $E_x[\overset{*}{T}_0]$, when $x \in [\lambda, V]$.

Let $\eta_\tau = \inf\{t \geq 0 : \overset{*}{I}_t \leq \tau\}$ and, for each $t \geq 0$,

$$\overset{*}{X}_t = \{\overset{*}{I}_t, t < \eta_\tau\}.$$

Note that, the state space of the process $\overset{*}{X}$ is the interval $(\tau, V]$, and let $\overset{*}{U}^\alpha$ be its α-potential. Starting at any $x \in [\lambda, V]$, $\eta_\tau = \overset{*}{T}_0$ almost surely, furthermore the sample paths of a spectrally positive Lévy process and a spectrally positive Lévy process reflected at its infimum behave the same way until they reach level τ, thus $\overset{*}{X}$ behaves the same way in both cases. It follows that, for each $\alpha \geq 0$, and $x \in [\lambda, V]$,

$$\boldsymbol{C}^\alpha{}_{g^*}(x) = \overset{*}{U}^\alpha g^*(x).$$

Note that the process N, is a spectrally positive Lévy process with the Lévy exponent $\phi_M(\theta) = \phi(\theta) + \theta M$, $\theta \geq 0$, where ϕ is the Laplace exponent of the process I. We denote the α-scale and adjoint α-scale functions, of the process N, by $W_M^{(\alpha)}$ and $Z_M^{(\alpha)}$, respectively.

The results (3.53), (3.54), and (3.55) below are analogous to (3.28), (3.29), and (3.30), respectively. Their proofs follow in a manner similar to the proof of these results, and observing that, for $\lambda \in [\lambda, V]$, $\overset{*}{T}_0 = \eta_\tau$ almost everywhere, P_x. We omit the proofs.

Theorem 3.24 For $\alpha > 0$, $\overset{*}{U}{}^{\alpha}$ is absolutely continuous with respect to the Lebesgue measure on $(\tau, V]$, and a version of its density is given by

$$\overset{*}{u}{}^{\alpha}(x, y) = \frac{Z_M^{(\alpha)}(V - x)W_M^{(\alpha)}(y - \tau)}{Z_M^{(\alpha)}(V - \tau)} - W_M^{(\alpha)}(y - x). \quad x, y \in (\tau, V] \tag{3.53}$$

Theorem 3.25 (i) Let $x \in [\lambda, V]$ and $\alpha \in R_+$. Then we have

$$E_x[e^{-\alpha \overset{*}{T}_0}] = \frac{Z_M^{(\alpha)}(V - x)}{Z_M^{(\alpha)}(V - \tau)}. \tag{3.54}$$

(ii) For $x \in [\lambda, V]$

$$E_x[\overset{*}{T}_0] = \overline{W}_M(V - \tau) - \overline{W}_M(V - x), \tag{3.55}$$

where, $\overline{W}_M(x) = \int_0^x W_M(y)dy$.

We now turn our attention to the case where $x < \lambda$. We first consider the case where the input process is a spectrally positive Lèvy process.

Theorem 3.26 Assume that the input process, I, is a spectrally positive Lévy process. For $x < \lambda$, and $z \geq \lambda$, we define

$$h_\alpha^*(x, dz) = \int_0^\lambda \upsilon(dz - y)u^\alpha(x, y)dy,$$

where $u^\alpha(x, y)$ is defined in (1.43), and denote $h_0^*(x, dz)$ by $h^*(x, dz)$. Then,

(i) For $\alpha \geq 0$, $x < \lambda$

$$E_x[e^{-\alpha \overset{*}{T}_0}] = \frac{1}{Z_M^{(\alpha)}(V - \tau)}\Big[\int_{[\lambda, V)} Z_M^{(\alpha)}(V - z)h_\alpha^*(x, dz) + \int_{[V, \infty)} h_\alpha^*(x, dz)\Big]. \tag{3.56}$$

(ii) for $\alpha \geq 0, x < \lambda$

$$
\begin{aligned}
E_x[\overset{*}{T}_0] &= \frac{W(\lambda - x)}{\eta(0)} - \overline{W}(\lambda - x) + \overline{W}_M(V - \tau) \\
&\quad - \int_\lambda^V \overline{W}_M(V - z) h^*(x, dz), \quad \text{if} \quad \eta(0) > 0, \\
&= \infty, \quad \text{if} \quad \eta(0) = 0
\end{aligned}
\tag{3.57}
$$

Proof The proof of (i) is similar to the proof of (3.33), with obvious modifications. The proof of (ii) follows in a manner similar to the proof of (3.34), using (3.40) and (3.55). We omit both proofs. ■

The following theorem gives results analogous to (3.56) and (3.57), when the input process is a spectrally positive Lèvy process reflected at its infimum.

Theorem 3.27 Assume that the input process, I, is a spectrally positive Lévy process reflected at its infimum. For $z \geq \lambda$, let $l_\alpha(dz)$, $L_\alpha(z)$, and $V_\alpha(\lambda)$ be as defined in (3.48), (3.49), and (3.50), respectively. Define

$$
g_\alpha(x, dz) = \begin{cases} \frac{l_\alpha(dz)}{W_+^{(\alpha)'}(\lambda)}, \; z > \lambda \\ \frac{V_\alpha(\lambda) - L_\alpha(\lambda)}{W_+^{(\alpha)'}(\lambda)} \delta_\lambda(dz). \end{cases}
$$

and denote $g_0(x, dz)$ by $g(x, dz)$.

Then,

(i) for $\alpha \geq 0, x < \lambda$

$$
E_x[e^{-\alpha \overset{*}{T}_0}] = \frac{1}{Z_M^{(\alpha)}(V - \tau)} [\int_{[\lambda, V)} Z_M^{(\alpha)}(V - z) g_\alpha(x, dz) + \int_{[V, \infty)} g_\alpha(x, dz)].
\tag{3.58}
$$

(ii) for $\alpha \geq 0, x < \lambda$

$$
\begin{aligned}
E_x[\overset{*}{T}_0] &= W(\lambda - x) \frac{W(\lambda)}{W_+'(\lambda)} - \overline{W}(\lambda - x) + \overline{W}_M(V - \tau) \\
&\quad - \int_{[\lambda, V)} \overline{W}_M(V - z) g(x, dz).
\end{aligned}
\tag{3.59}
$$

Proof The proof of (i) follows in a manner similar to the proof of (3.33), using (3.51), (3.52), and (3.54). The proof of (ii) is similar to the proof of (3.34), using (3.47) and (3.55), we omit the details. ■

3.6 Examples

In this section, we give formulas for the basic entities needed to compute the total discounted as well as the long-run average costs for specific cases of the input process.

Example 1 Assume that the input process I is an increasing compound Poisson process with zero drift, intensity u, and F as the distribution function of the size of each jump. It follows that the Laplace exponent is of the form,

$$\psi(\alpha) = u \int_0^\infty (1 - e^{-\alpha x}) F(dx),$$

and $E_0(I_1) = u\mu$, where μ is the expected jump size of the compound Poisson process.

Define, for any $\alpha \geq 0$ and $y \geq 0$, $F_\alpha(x) = \frac{u}{u+\alpha} F(x)$. For $n \in N_+$, we let $F_\alpha^{(n)}(x)$ be the nth convolution of $F_\alpha(x)$, where $F_\alpha^{(0)}$ is the Dirac measure at $\{0\}$. Suppose that M_α is the renewal function corresponding F_α. That is to say, for $x \in R_+$,

$$M_\alpha(x) = \sum_{n=0}^{\infty} F_\alpha^{(n)}(x).$$

From (1.34) we have, for all $y \geq 0$. For each $y \geq 0$, the α-potential is as follows:

$$\mathbf{R}^\alpha(dy) = \frac{1}{u+\alpha} M_\alpha(dy) \tag{3.60}$$

From (3.18) and (3.19) we have, for $x \leq \lambda$,

$$E_x[\exp(-\alpha \hat{T}_0)] = 1 - \frac{\alpha}{u+\alpha} M_\alpha(\lambda - x), \tag{3.61}$$

and

$$E_x[\hat{T}_0] = \frac{1}{u} M(\lambda - x), \tag{3.62}$$

where M is the renewal function corresponding to F.

Also, from (3.23) we have, for $x \leq \lambda$

$$E_x[e^{-\alpha \hat{T}_0}, I_{\hat{T}_0} \in dz] = \frac{u}{u+\alpha} [\int_{(0,\lambda]} F(dz - y) M_\alpha(dy - x)].$$

Letting $\alpha \to 0$, in both sides of the last equation above, the distribution function of $I_{\hat{T}_0}$ is given as follows:

$$G(dz) = [\int_{(0,\lambda]} F(dz - y)M(dy - x)]I_{[\lambda,\infty)}(z). \tag{3.63}$$

Let $\bar{F} = 1 - F$. Assuming that $\mu < \infty$, we define the probability density function $h(x) = \frac{\bar{F}(x)}{\mu}$, and let H be the distribution function associated with h. Since the process N, is of bounded variation, assuming that $\rho = \frac{u\mu}{M} < 1$, from (1.25), it follows that the scale function of the process N is as follows:

$$W_M(x) = \frac{1}{M}\sum_{n=0}^{\infty} \rho^n H^{(n)}(x). \tag{3.64}$$

Using (3.23), (3.33), (3.34), and (3.60)–(3.63) the total discounted as well as the long-run average costs given in (3.11) and (3.12) are computed.

Example 2 Assume that the input process, I, is an inverse Brownian motion with no drift. From (1.5) it follows that, for $\theta \geq 0$, $\mu > 0$, the Laplace exponent of this process is as follows

$$\psi(\theta) = \frac{1}{\sigma^2}(\sqrt{2\theta\sigma^2 + \mu^2} - \mu).$$

Furthermore, its Lévy measure is of the form

$$\nu(dy) = \frac{1}{\sigma\sqrt{2\pi y 3}} e^{-(y\mu^2/2\sigma^2)},$$

and $E_0(I_1) = \frac{1}{\mu} < \infty$. For $x \geqslant 0$, we let $\bar{\nu}(x) = \int_{[x,\infty)} \nu(dy)$.

Substituting this Lévy component above in (1.10) it is seen that: (we omit the proof)

$$\eta(\alpha) = \frac{\alpha}{M} + \frac{(1 - M\mu) + \sqrt{2\alpha M\sigma^2 + (1 - M\mu)^2}}{M^2\sigma^2}. \tag{3.65}$$

For $\alpha \geq 0$, suppose that $\mathbf{R}^{\alpha}$ is the α-potential of the process I. Let $\varphi_Z(.)$ be the standard normal density function, erf() and erf $c()$ be the well-known error and complimentary error functions, respectively. From Corollary 1.20, it follows that $\mathbf{R}^{\alpha}$ is absolutely continuous with respect to the Lebesgue measure and a version of its density is given in (1.35).

For $x \leq \lambda$, we have

$$
\begin{aligned}
E_x[\exp(-\alpha \hat{T}_0)] &= \alpha \mathbf{R}^\alpha \mathbf{I}_{[\lambda-x,\infty)}(0). \\
&= \frac{\alpha\sigma^2-\mu}{\alpha\sigma^2-2\mu} e^{\alpha(\lambda-x)(\frac{\alpha\sigma^2}{2}-\mu)} \operatorname{erf} c(\sqrt{\lambda-x}\frac{\alpha\sigma^2-\mu}{\sqrt{2\sigma^2}}) \\
&- \frac{\mu}{\alpha\sigma^2-2\mu} \operatorname{erf} c(\frac{\sqrt{\lambda-x}\mu}{\sqrt{2\sigma^2}}),
\end{aligned} \tag{3.66}
$$

where the first equation follows from (3.18), and the last equation follows from (1.35) by integrating $\mathbf{R}^\alpha(dy)$ over the interval $[\lambda-x,\infty)$.

Inverting the right-hand side of (3.66), with respect to α, it follows that, given $Z_0 = x \le \lambda$, the distribution function of $\hat{T}_0$ (denoted by $F_{\hat{T}_0}()$) is given by

$$
F_{\hat{T}_0}(t) = \frac{1}{2} \operatorname{erf} c\{\frac{(\lambda-x)\mu-t}{\sqrt{2\sigma^2}}\} - \frac{1}{2} e^{2\mu t/\sigma^2} \operatorname{erf} c\{\frac{(\lambda-x)\mu+t}{\sqrt{2\sigma^2}}\}. \ t \ge 0. \tag{3.67}
$$

Furthermore, for $x \le \lambda$

$$
\begin{aligned}
E_x[\hat{T}_0] &= \mathbf{R}\mathbf{I}_{[0,\lambda)}(x) \\
&= \sigma \int_0^{\lambda-x} \frac{1}{\sqrt{y}} \varphi_Z(\sqrt{y}\frac{\mu}{\sigma}) dy + \frac{\mu}{2} \int_0^{\lambda-x} \operatorname{erf} c(-\sqrt{\frac{y}{2}}\frac{\mu}{\sigma})\, dy. \\
&= \frac{(\lambda-x)\mu}{2} + \sigma\sqrt{\lambda-x}\,\varphi_Z(\sqrt{\lambda-x}\frac{\mu}{\sigma}) \\
&+ \frac{(\lambda-x)\mu^2+\sigma^2}{2\mu} \operatorname{erf}(\sqrt{\frac{\lambda-x}{2}}\frac{\mu}{\sigma}),
\end{aligned} \tag{3.68}
$$

where the first equation follows from (3.19), the second equation follows from (1.35), and the third equation follows from the second equation upon tedious calculations which we omit.

We now turn our attention to computing the distribution function of $I_{\hat{T}_0}$ (denoted by $F_{I_{\hat{T}_0}}()$). We first need the following identity which expresses the Lévy exponent $\psi(\alpha)$ in a form suitable for computing $F_{I_{\hat{T}_0}}$. The proof of this identity follows after some simple algebraic manipulations which we omit.

$$
\begin{aligned}
\psi(\theta) &= \frac{\sqrt{2\theta\sigma^2+\mu^2}-\mu}{\sigma^2} \\
&= \frac{2}{\sigma^2}[\frac{\theta}{\psi(\theta)} - \mu].
\end{aligned} \tag{3.69}
$$

For each $\theta \in R_+$, we write

$$\int_{\lambda}^{\infty} e^{-\theta x} F_{I_{\hat{T}_0}}(x)dx = \frac{\psi(\theta)}{\theta}\int_{\lambda}^{\infty} e^{-\theta x} r(x)dx$$
$$= \frac{2}{\sigma^2}[\frac{1}{\psi(\theta)}\int_{\lambda}^{\infty} e^{-\theta x} r(x)dx - \frac{\mu}{\theta}\int_{\lambda}^{\infty} e^{-\theta x} r(x)dx]$$
$$= \frac{2}{\sigma^2}[\int_{0}^{\infty} e^{-\theta x} r(x)dx \int_{\lambda}^{\infty} e^{-\beta x} r(x)dx - \frac{\mu}{\theta}\int_{\lambda}^{\infty} e^{-\theta x} r(x)dx]$$
$$= \frac{2}{\sigma^2}[\int_{\lambda}^{\infty} e^{-\theta x}\{\int_{\lambda}^{x}(r(x-y)-\mu)r(y)dy\}dx],$$

where the first equation follows from (3.21), the second equation follows from (3.69), the third equation follows since $\int_0^{\infty} e^{-\theta x} r(x)dx = \frac{1}{\psi(\theta)}$, and the fourth equation follows from the third equation through integration by parts.

From the last equation above it follows that, for each $x \geq \lambda$

$$F_{I_{\hat{T}_0}}(x) = \frac{2}{\sigma^2}[\int_{\lambda}^{x}\{r(x-y)-\mu\}r(y)dy]. \tag{3.70}$$

The computation of $E_x[e^{-\alpha\hat{T}_0}, I_{\hat{T}_0} \in dz]$ is done using (3.23) and (1.35).

Since the process N is spectrally positive and of bounded variation, $E[I_1] = \frac{1}{\mu} < \infty$, and assuming that $\frac{1}{\mu} < M$, then from (1.25), the scale function is as follows:

$$W_M(x) = \frac{1}{M}\sum_{n=0}^{\infty}\rho^n F^{(n)}(x),$$

where $\rho = \frac{1}{M\mu}$, F is the distribution function corresponding to the density $f(x) = \mu\bar{\nu}(x), x > 0$. For $\alpha > 0$, the α-scale function is computed using (1.13). For the finite capacity dam, the computations of $E_x[e^{-\alpha T_0^*}]$ and $E_x[T_0^*]$, are established using (3.33) and (3.34), respectively. For infinite dams, they are computed using (3.35) and (3.36), respectively.

Substituting (1.35), (3.28), (3.66), (3.68), $E_x[e^{-T_0^*}]$, and $E_x[T_0^*]$ into (3.13)–(3.15), we obtain the values of $C_\alpha(x)$ and $C(x)$. The total discounted cost can be determined explicitly by substituting the values of $C_\alpha(x)$, $E_x[e^{-T_0^*}]$ in (3.11). Finally, The long-run average cost is computed by substituting the values of $C(x)$, and $E_x[T_0^*]$ into (3.12).

Example 3 Assume that the input process is a spectrally positive Lévy process of bounded variation with Laplace exponent described in (1.14), reflected at its infimum. Let υ be its Lévy measure, $\bar{\nu}()$ be as defined in Example 2 above, $\mu = \int_0^{\infty} x\upsilon(dx)$, and assume that $0 < \mu < \infty$. For every $x \in R_+$, we define the probability density function $f(x) = \frac{\bar{\upsilon}(x)}{\mu}$, and $F(x)$ is the distribution function corresponding to f. Define $\rho = \frac{\mu}{\varsigma}$ and assume that $\rho < 1$. From (1.25), the scale function of the underlying process, is of the form

$$W(x) = \frac{1}{\varsigma}\sum_{n=0}^{\infty}\rho^n F^{(n)}(x).$$

For $\alpha > 0$, $W^{(\alpha)}$ is computed using (1.13).

Let $\varsigma^* = \varsigma + M$, $\overset{*}{\rho} = \frac{\mu}{\varsigma^*}$, and assume that $\rho^* < 1$, it follows that the scale function of the process N is of the form

$$W_M(x) = \frac{1}{\varsigma^*}\sum_{n=0}^{\infty}\overset{*n}{\rho} F^{(n)}(x).$$

The values of $E_x[\exp(-\alpha\hat{T}_0)]$, and $E_x[\hat{T}_0]$ are computed using (3.46) and (3.47), respectively. While, the computation of $E_x[e^{-\alpha\hat{T}_0}, I_{\hat{T}_0} \in dz]$, is established using (3.51) and (3.52).

The values of the functions $C_g^\alpha(x)$, and $C_g(x)$ are computed using (3.44), while $\mathbf{C}^\alpha{}_{g^*}(x)$ and $\mathbf{C}_{g^*}(x)$ are computed using (3.53), and the computations of $E_x[e^{-T_0^*}]$, $E_x[T_0^*]$, are established using (3.58) and (3.59). The values of $C_\alpha(x)$ and $C(x)$ are determined using (3.13)–(3.15). Finally the total discounted and the long-run average costs are determined using (3.11) and (3.12), respectively.

Remark 3 (a) Assume that the input process is a compound Poisson process with a negative drift. In this case, $\upsilon(dx) = \lambda G(dx)$, where $\lambda > 0$ and G is a distribution function of a positive random variable $[0, \infty)$, describing the size of each jump of the compound Poisson process. In this case, $f(x) = \frac{\overline{G}(x)}{m}$ and $\rho = \frac{\lambda m}{\varsigma}$, where $\overline{G} = 1 - G$ and $m = \int_0^\infty \overline{G}(x)dx$, which is assumed to be finite

(b) Assume that the input process is a gamma process with negative drift. The Lévy measure is given by $\upsilon(dx) = \alpha\frac{e^{-x/\beta}}{x}dx$, $\alpha, \beta > 0$, and its Laplace exponent is given (1.26). In this case, $E(I_1) = \alpha\beta - \varsigma$, and $\rho = \frac{\alpha\beta}{\varsigma}$. It follows that $f(x) = \frac{1}{\beta}\int_x^\infty \frac{e^{-y/\beta}}{y}$, we note that the right-hand side is denoted by $E_1(x)$ in p. 227 of [14]. Direct integrations yields, $F(x) = (1 - e^{-x/\beta}) + xf(x)$.

(c) Assume that the input process is an inverse Gaussian process with a negative drift, and with Lévy measure is given by $\upsilon(dx) = \frac{1}{\sigma\sqrt{2\pi x^3}}e^{-xc^2/2\sigma^2}$, $\sigma, c > 0$. It follows that $E(I_1) = \frac{1}{c} - \varsigma$, and $\rho = \frac{1}{c\varsigma}$. In this case, $f(x) = c\int_x^\infty \upsilon(dy)$, and $F(x) = \mathrm{erf}(c\sqrt{x/2\sigma^2}) + xf(x)$.

Example 4 Assume that the input process is a Brownian motion with drift term $\mu \in R$, variance term σ^2, reflected at its infimum. In this case, the Lévy measure $\nu = 0$, and from (1.9) we have, that for $\theta \geq 0$, $\phi(\theta) = -\mu\theta + \frac{\theta^2\sigma^2}{2}$. It follows that, for $\alpha \geq 0$, $\eta(\alpha) = \frac{\sqrt{2\alpha\sigma^2+\mu^2}+\mu}{\sigma^2}$. Let $\delta = \sqrt{2\alpha\sigma^2 + \mu^2}$, from (1.19), we have,

$$W^{\alpha}(x) = \frac{2}{\delta} e^{\mu x/\sigma^2} \sinh(\frac{x\delta}{\sigma^2}),$$
$$Z^{\alpha}(x) = e^{\mu x/\sigma^2} \left(\cosh(\frac{x\delta}{\sigma^2}) - \frac{\mu}{\delta} \sinh(\frac{x\delta}{\sigma^2}) \right).$$

Note that $W^{\alpha}(x)$ is differentiable, and $W^{\alpha'}(x) = \frac{\mu}{\sigma^2} W^{\alpha}(x) + \frac{2}{\sigma^2} e^{\mu x/\sigma^2} \cosh(\frac{x\delta}{\sigma^2})$. It follows that $\frac{W^{(\alpha)}(\lambda)}{W^{(\alpha)'}(\lambda)} = \left(\frac{\sigma^2}{\mu+\delta \coth(\frac{\lambda\delta}{\sigma^2})}\right)$. Substituting the values of $Z^{(\alpha)}(\lambda - x)$, $W^{(\alpha)}(\lambda - x)$, and $\frac{W^{(\alpha)}(\lambda)}{W^{(\alpha)'}(\lambda)}$ in (3.46), we have, for $\alpha \geq 0, x \leq \lambda$

$$E_x[e^{-\alpha \hat{T}_0}] = e^{\mu(\lambda - x)} \left[\cosh\left(\frac{(\lambda - x)\delta}{\sigma^2}\right) - \frac{1}{\delta} \sinh\left(\frac{(\lambda - x)\delta}{\sigma^2}\right) \left(\mu + \frac{2\alpha\sigma^2}{\mu + \delta \coth(\frac{\lambda\delta}{\sigma^2})}\right)\right]. \tag{3.71}$$

(i) Assume that $\mu \# 0$. It follows that, for $x \geq 0$, $W(x) = \frac{e^{2\mu x/\sigma^2} - 1}{\mu}$, $W'(x) = \frac{2e^{2\mu x/\sigma^2}}{\sigma^2}$ and $\overline{W}(x) = \frac{\sigma^2}{2\mu^2}(e^{2\mu x/\sigma^2} - 1) - \frac{x}{\mu}$. Substituting the values of $W(\lambda - x)$, $\overline{W}(\lambda - x)$, $W(\lambda)$, and $W'(\lambda)$ in (3.47) we have, for $x \leq \lambda$,

$$E_x[\hat{T}_0] = \frac{\lambda - x}{\mu} + \frac{\sigma^2}{2\mu^2}\left[e^{-2\mu\lambda/\sigma^2} - e^{-2\mu x/\sigma^2}\right]. \tag{3.72}$$

We note that, $W^{(\alpha)}(0) = 0$, $\overset{(2)}{U^{\alpha}}$ in (3.44) is absolutely continuous with respect to the Lebesgue measure on $[0, \lambda)$ and for $y \in [0, \lambda)$, $W^{(\alpha)}(dy) = W^{\alpha'}(y)dy$. Substituting the values of $W^{(\alpha)}(\lambda - x)$, $W^{\alpha'}(\lambda)$, $W^{(\alpha)}(y - x)$, and $W^{\alpha'}(y)$ in (3.44), we get a version of the density of $\overset{(2)}{U^{\alpha}}$. Thus, $C_g^{\alpha}(x) = \overset{(2)}{U^{\alpha}} g(x)$, and $C_g(x) = \overset{(2)}{U} g(x)$ are computed.

Let $\overset{*}{\mu} = \mu - M$, $\overset{*}{\delta} = \sqrt{2\alpha\sigma^2 + \overset{*}{\mu}^2}$, note that the scale function of the process N is of the form $W_M^{(\alpha)}(x) = \frac{2}{\delta} e^{\overset{*}{\mu} x/\sigma^2} \sinh(\frac{x\overset{*}{\delta}}{\sigma^2})$. Since the input process is continuous, $I_{\hat{T}_0} = \lambda < V$, almost surely. Therefore, the terms $\boldsymbol{C}^{\alpha}{}_{g^*}((I_{\hat{T}_0} \wedge V)$, $\boldsymbol{C}_{g^*}(I_{I_{\hat{T}_0}} \wedge V)$, in (3.14) and (3.15) reduce to $\boldsymbol{C}^{\alpha}{}_{g^*}(\lambda)$, $\boldsymbol{C}_{g^*}(\lambda)$, respectively, almost surely. Furthermore,

$$E_x[e^{-\alpha \overset{*}{T}_0}] = E_x[e^{-\alpha \hat{T}_0}] E_{\lambda}[e^{-\alpha \overset{*}{T}_0}], \tag{3.73}$$

where $E_x[e^{-\alpha \hat{T}_0}]$ is given in (3.71) and $E_{\lambda}[e^{-\alpha \overset{*}{T}_0}]$ is given in (3.54).

Let $\lambda^* = V - \lambda$ and $\tau^* = V - \tau$, then

$$E_{\tau}[\overset{*}{T}_0] = E_{\tau}[\hat{T}_0] + E_{\lambda}[\overset{*}{T}_0], \tag{3.74}$$

Note that

$$E_\lambda[\overset{*}{T}_0] = \frac{\lambda^* - \tau^*}{\mu^*} + \frac{\sigma^2}{2\overset{*}{\mu}^2}\left[e^{2\overset{*}{\mu}\tau^*/\sigma^2} - e^{2\mu\lambda^*/\sigma^2}\right], \tag{3.75}$$

where the last equation follows from (3.55) after some tedious calculations which we omit.

Substituting (3.71)–(3.73), and the values of $C_g^\alpha(x)$, $\boldsymbol{C}^\alpha{}_{g^*}(\lambda)$ in (3.13) and (3.14), $C_\alpha(x)$ is computed. The total discounted cost is obtained by substituting the values of $C_\alpha(x)$, $C_\alpha(\tau)$, and $E_x[e^{-\alpha\overset{*}{T}_0}]$ in (3.11). Finally, the long-run average cost is computed by substituting the values of $C(\tau)$ and $E_\tau[\overset{*}{T}_0]$ in (3.12).
(ii) Consider the case when $\mu = 0$. In this case, $\delta = \sqrt{2\alpha\sigma^2}$, letting $\mu \to 0$, in the corresponding equations above, we have

$$E_x[e^{-\alpha\hat{T}_0}] = \left[\cosh((\lambda - x)\delta/\sigma^2) - \sigma^2\frac{\sinh\left((\lambda - x)\delta/\sigma^2\right)}{\coth(\frac{\lambda\delta}{\sigma^2})}\right], \tag{3.76}$$

$$E_x[\hat{T}_0] = \frac{\lambda^2 - x^2}{\sigma^2}, \tag{3.77}$$

and $E_\lambda[\overset{*}{T}_0]$ is obtained by replacing μ^* by $-M$ in (3.75).

The values of the total discounted and the long-run average costs are computed in manners similar to those used in (i), with obvious modifications.

Example 5 Assume that the input process is a Brownian motion with drift term $\mu > 0$ and variance parameter σ^2. Substituting the values of $W^{(\alpha)}(x)$, $Z^{(\alpha)}(x)$, given in Example 4, in (3.39) we have, for $x \le \lambda$, $E_x[e^{-\alpha\hat{T}_0}] = \exp((\delta - \mu)(x - \lambda))$. Substituting $\frac{2\mu}{\sigma^2}$ for $\eta(0)$, $\frac{1}{\mu}(e^{2\mu x/\sigma^2} - 1)$ and $\frac{\sigma^2}{2\mu^2}(e^{2\mu x/\sigma^2} - 1) - \frac{x}{\mu}$ for $W(x)$ and $\overline{W}(x)$, respectively, in (3.40) we have, for $x \le \lambda$, $E_x[\hat{T}_0] = \frac{\lambda - x}{\mu}$. The computations of the other entities in the cost functionals (3.11) and (3.12) can be obtained in a manner similar to those discusses in Example 4, with obvious modifications.

Example 6 Assume that the input process is a spectrally positive process of bounded variation with a stable subordinator, reflected at its infimum. For $\beta \in (0, 1)$, from (1.28), we know that the scale function is a Mittag-Leffler function of the form

$$W(x) = \frac{1}{\varsigma}\sum_{k=0}^{\infty}\frac{(x^{(1-\beta)}/\varsigma)^k}{\Gamma(1 + k(1 - \beta))}.$$

It follows that

$$\overline{W}(x) = \frac{x}{\varsigma}\sum_{k=0}^{\infty}\frac{(x^{(1-\beta)}/\varsigma)^k}{\Gamma(2 + k(1 - \beta))},$$

and

$$W^{'}(x) = \frac{1}{\varsigma x}\sum_{k=1}^{\infty}\frac{(x^{(1-\beta)}/\varsigma)^k}{\Gamma(k(1-\beta))}.$$

Furthermore, with $\varsigma^* = \varsigma + M$, we have

$$W_M(x) = \frac{1}{\varsigma^*}\sum_{k=0}^{\infty}\frac{(x^{(1-\beta)}/\varsigma^*)^k}{\Gamma(1+k(1-\beta))},$$

$$\bar{W}(x) = \frac{x}{\varsigma^*}\sum_{k=0}^{\infty}\frac{(x^{(1-\beta)}/\varsigma^*)^k}{\Gamma(2+k(1-\beta))},$$

and

$$W_M^{'}(x) = \frac{1}{\varsigma^* x}\sum_{k=1}^{\infty}\frac{(x^{(1-\beta)}/\varsigma^*)^k}{\Gamma(k(1-\beta))}.$$

Using the above equations and (1.13), the computations of the different entities involved in the total discounted and long-run average costs are established using the corresponding formulas for these identities as given in Sect. 3.5.

Example 7 Assume that the input process, I, is a gamma subordinator with no drift term and parameters α, β as given in (1.6). From (33) of [15], for $x \leq \lambda$, we have

$$E_x[\hat{T}_0] = (\lambda - x) + \frac{1}{2} - E[\frac{\exp(-(\lambda - x)(Z+1))(Z+1)}{(\ln(Z))^2 + \pi^2}],$$

where Z is a Pareto random variable with density function

$$f(x) = \frac{1}{(1+x)^2}\mathbf{I}_{[0,\infty)}(\mathbf{x}).$$

If the dam has very large capacity ($V = \infty$) and the penalty costs functions are constant, i.e., $g(x) = c_1$ and $g^*(x) = c_2$. Assume that $M > \alpha\beta$, from (3.38), it follows that the long-run average cost function is computed as follows:

$$\mathbb{C}(\lambda, \tau) = \frac{M^* K}{E_\tau[\hat{T}_0]} + \frac{1}{M}(c_1 M^* + \alpha\beta c_2) - \alpha\beta R,$$

where $M^* = M - \alpha\beta$.

References

1. Lam Y, Lou JH (1987) Optimal control of a finite dam: Wiener process input. J Appl Prob 35:482–488
2. Attia F (1987) The control of a finite dam with penalty cost function; Wiener process input. Stoch Process Appl 25:289–299
3. Lee EY, Ahn SK (1998) P^M_λ policy for a dam with input formed by a compound Poisson process. J Appl Prob 24:186–199
4. Bae J, Kim S, Lee EY (2002) A P^M_λ policy for an M/G/1 queueing system. Appl Math Model 26:929–939
5. Bae J, Kim S, Lee EY (2003) Average cost under $P^M_{\lambda,\tau}$-policy in a finite dam with compound Poisson input. J Appl Prob 40:519–526
6. Alili L, Kyprianou AE (2005) Some remarks on the first passage of Lévy processes, the American put and pasting principles. Ann Appl Probab 15:2062–2080
7. Abdel-Hameed M (2000) Optimal control of a dam using $P^M_{\lambda,\tau}$ policies and penalty cost when the input process is a compound Poisson process with positive drift. J Appl Prob 37:408–416
8. Abdel-Hameed M (2011) Control of dams using $P^M_{\lambda,\tau}$ policies when the input process is a nonnegative Lévy process. Int J Stoch Anal. Article ID 916952
9. Abdel-Hameed M (2012) Control of dams when the input process is either spectrally positive Lévy or spectrally positive Lévy reflected at its infimum. arxiv:1208.6559v1, to appear
10. Miller BM, McInnes DJ (2011) Management of a large dam via optimal Price control. In: International Federation of Automatic Control, Milano, pp 12432–12438
11. Ross SM (1983) Stochastic processes. Wiley, New York
12. Kyprianou AE (2006) Introductory lecture notes on fluctuations of Lévy processes with applications. Springer, Berlin
13. Zhou XW (2004) Some fluctuation identities for Lévy process with jumps of the same sign. J Appl Prob 41:1191–1198
14. Abramowitz M, Stegun IA (1964) Handbook of mathematical functions. Dover, New York
15. Frenk J, Nicolai R (2007) Approximating the randomized hitting time distribution of a nonstationary gamma process. Econometric report 2007-18. Econometric Institute and ERIM, Erasmus University
16. Zuckerman D (1977) Two-stage output procedure for a finite dam. J Appl Prob 14:421–425

Appendix
Preliminaries

1. Doob's Optional Sampling Theorem

For any stochastic process Y and $t \in R_+$ we let $\mathcal{F}_t = \sigma(Y_s, s \leq t)$.

Definition 1 A process $K = \{K_t, t \in R_+\}$ is adapted with respect to $\mathcal{F}_t$, if for every $t \in R_+$, $K_t \in \mathcal{F}_t$.

Definition 2 A process $K = \{K_t, t \in R_+\}$ is predictable with respect to $\mathcal{F}_t$, if for every $t \in R_+$, $K_t \in \mathcal{F}_{t-}$.

Definition 3 A stochastic process $Y = \{Y_t, t \geq 0\}$ is a martingale if for all $s, t \in R_+$

$$E[Y_{t+s} \mid \mathcal{F}_t] = Y_t,$$

almost everywhere **P**.

Definition 4 Let $Y = \{Y_t, t \geq 0\}$ be a stochastic process. A positive random variable $T : \Omega \to \overline{R}_+$ is a stopping time with respect to $\mathcal{F}_\infty$ if for each $t \in R_+$, the event $\{T \leq t\} \in \mathcal{F}_t$.

Remark (a) If a random variable T is discrete with support $\{s_1 < s_2 < \cdot\}$, then T is a stopping time with respect to $\mathcal{F}_\infty$ if and only if for every $n = 1, 2, ..$, the event $\{T = s_n\} \in \mathcal{F}_{s_n}$.

(b) Note that the constant random variable is a stopping time. This is true because if T is such a random variable, then for each $t \in R_+$, $\{T \leq t\}$ is either Ω or the null set, both of which are in $\mathcal{F}_t$.

(c) If T_1 and T_2 are two stopping times, then $T_1 \wedge T_2$ is a stopping time as well. This is true since, for every $t \in R_+$, the events $\{T_1 > t\}$ and $\{T_2 > t\} \in \mathcal{F}_t$. But $\{T_1 \wedge T_2 > t\} = \{T_1 > t\} \cap \{T_2 > t\}$, hence $\{T_1 \wedge T_2 > t\} \in \mathcal{F}_t$ and since $\mathcal{F}_t$ is a sigma algebra, then the event $\{T_1 \wedge T_2 \leq t\} \in \mathcal{F}_t$.

The following is known as the *Doob's Optional Sampling Theorem*.

M. Abdel-Hameed, *Lévy Processes and Their Applications in Reliability and Storage*, SpringerBriefs in Statistics, DOI: 10.1007/978-3-642-40075-9, © The Author(s) 2014

Theorem 5 Let Y be a bounded martingale, suppose that T is a finite stopping time. Then $E[Y_T] = E[Y_0]$.

2. Markov and Strong Markov processes

Definition 6 A stochastic process $Y = \{Y_t, t \geq 0\}$ is said to be a *Markov* process if, for $s, t \in R_+$ and every bounded measurable function $f : \Omega \to R$,

$$E[f(Y_{t+s}) \mid F_t] = E[f(Y_{t+s}) \mid \sigma(Y_t)].$$

Loosely speaking, a stochastic process is a Markov process, if the future is independent of the past given the present. All processes dealt with in this monograph are Markovian. A stronger version of the Markov property is the Strong Markov property. It is satisfied if the above equality holds when the fixed time t is replaced by a stopping time. Formally,

Definition 7 A stochastic process $Y = \{Y_t, t \geq 0\}$ is said to be a *strong Markov* process if, for $s \in R_+$, every bounded measurable function $f : \Omega \to R$, and every stopping time T

$$E[f(Y_{T+s}) \mid F_T] = E[f(Y_{T+s}) \mid \sigma(X_T)], \text{ on } \{T < \infty\}$$

3. The Monotone Class Theorems

Definition 8 Let S be set and Ş be a collection of subsets of S. The class Ş is called a $\pi-$system if it is closed under finite intersections.

Theorem 9 (The monotone Class Theorem 1). Let S be a set and Ş is a $\pi-$system on S. Let $\mathcal{F}$ be a vector space of bounded real-valued functions on S such that

(i) $A \in$Ş implies $\mathbf{I}_A \in \mathcal{F}$,
(ii) $\mathcal{F}$ contains the constant functions, and
(iii) if $(f_n) \subset \mathcal{F}$ is a sequence of positive increasing functions with $\sup_n \sup_\omega \mid f_n(\omega) \mid < \infty$, then $f = \lim_n f_n \in \mathcal{F}$.

Then, $\mathcal{F}$ contains every bounded real-valued σ(Ş) measurable function on S.

The following is an another version of the above theorem that is useful in many applications.

Theorem 10 (The monotone Class Theorem 2). Let S be a set. Let ₺ be a collection of bounded real-valued functions on S that is closed under the formation of products (i.e., if $f, g \in$ ₺, then $fg \in$ ₺), and let σ(₺) be the sigma algebra generated by ₺. Let $\mathcal{F} \supset$ ₺ be a vector space of bounded real-valued functions on Ω such that

(i) $\mathcal{F}$ contains the constant functions, and
(ii) if $(f_n) \subset \mathcal{F}$ is a sequence of positive increasing functions with $\sup_n \sup_\omega | f_n(\omega) | < \infty$, then $f = \lim_n f_n \in \mathcal{F}$.

Then, $\mathcal{F}$ contains every bounded real-valued $\sigma(\pm)$ measurable function on S.

4. Poisson Random Measure

Definition 11 Let $(K, \aleph)$ be a measurable space. A mapping $M : \aleph \to \overline{R}_+$ is said to be a measure on $\aleph$ if $M(\bigcup_n B_n) = \sum_n M(B_n)$, for all disjoint sets $B_1, B_1, \ldots$ in $\aleph$.

Definition 12 Let $(\Omega, F, \mathbf{P})$ be a probability space and $(K, \aleph)$ be a measurable space. A mapping $M : (\Omega \times \aleph) \to \overline{R}_+$ is said to be a random measure on $(K, \aleph)$ provided that

(i) $\omega \to M(\omega, B)$ is a random variable for each $B \in \aleph$, and
(ii) $B \to M(\omega, B)$ is a measure on $(K, \aleph)$ or each $\omega \in \Omega$.

Definition 13 A random measure M on $(\Omega \times (R_+ \times R_0))$ is said to be a *Poisson random measure* on $(R_+ \times R_0)$ with mean measure m provided that

(i) For each $B \in \sigma(R_+ \times R_0)$, the random variable $M(B, \omega)$ has a Poisson distribution with range $\overline{N}_+$, where $M(B) = \infty$ almost everywhere, if and only if $m(B) = \infty$.
(ii) For any disjoint sets $B_1, \ldots, B_n$, in $\sigma(R_+ \times R_0)$, the random variables $M(B_1), \ldots, M(B_n)$ are independent,

Adopting the standard terminology in probability, we will suppress the ω in $M(\omega, B)$ from now on.

5. Markov Renewal and Semi-Markov Processes

Let $(\Omega, F, \mathbf{P})$ be a probability space, and E be an arbitrary set. For each $n \in N_+$, we let

$$X_n : \Omega \to E,$$

$$T_n : \Omega \to R_+,$$

$F_n = \sigma\{(X_k, T_k), k \leq n\}$, and assume that

$$0 = T_0 \leq T_1 \leq T_2 \leq \cdot.$$

Definition 14 The process $(X, T) = \{(X_n, T_n), n \in N_+\}$ is called a Markov renewal process with state space E if, for each $n \in N_+,\ t \in R_+$, and any Borel set $A \subset E$,

$$P\{X_{n+1} \in A, T_{n+1} - T_n \leq t \mid \mathcal{F}_n\} = P\{X_{n+1} \in A, T_{n+1} - T_n \leq t \mid X_n\}.$$

We will assume that the process (X, T) is homogeneous, that is for each $n \in N_+,\ t \in R_+$, and any Borel set $A \subset E$,

$$P\{X_{n+1} \in A, T_{n+1} - T_n \leq t \mid X_n\} = P\{X_1 \in A, T_1 \leq t \mid X_0\}.$$

Definition 15 For each $x \in E,\ t \in R_+$, and any Borel set $A \subset E$, we write

$$Q(x, A, t) = P\{X_1 \in A, T_1 \leq t \mid X_0 = x\}$$

Then, $Q(x, A, t)$ is called the semi-Markov kernel of the Markov renewal process (X, T).

Definition 16 For $t \in R_+$, let

$$Y_t = \{X_n, T_n \leq t < T_{n+1}\}.$$

The process $Y = \{Y_t, t \in R_+\}$ is called the semi-Markov process associated with the Markov renewal process (X, T).

Definition 17 Suppose that, for each $x \in E,\ t \in R_+$, and any Borel set $A \subset E$,

$$Q^n(x, A, t) = P\{X_n \in A, T_n \leq t \mid X_0 = x\},$$
$$Q^0(x, A, t) = \mathbf{I}_{\{x \in A\}}.$$

Let

$$R(x, A, t) = \sum_{n=0}^{\infty} Q^n(x, A, t).$$

Then, $R(x, A, t)$ is called the Markov renewal kernel corresponding to Q.

Definition 18 If the state space E consists of a single point, then $\{T_n, n \in N_+; T_0 = 0\}$ forms a renewal process. That is if, for $n \geq 1$, we let $W_n = T_n - T_{n-1}$, then $\{W_n, n = 1, 2, \ldots\}$ is a sequence of independent identically distributed random variables. Let F be the distribution function of the random variable W_1, for $n \geq 1$, let $F^{(n)}$ be the nth convolution of F. For $x \in R_+$, the renewal function R is defined as follows:

$$R(x) = \sum_{n=0}^{\infty} F^{(n)}(x),$$

where, $F^{(0)}(x) = 1$, for all $x \in R_+$.

References

1. Blumenthal RM (1992) Excursions of Markov processes. Birkhauser, Boston
2. Blumenthal RM, Getoor RK (1968) Markov processes and potential theory. Academic Press, New York
3. Cinlar E (1975) Introduction to Stochastic processes. Prentice Hall, Englewood Cliffs
4. Dellacherie C, Meyer PA (1993) Probability and potential. Herman, Paris

Index